I WILL TELL YOU ALGEBRA STORIES YOU'VE NEVER HEARD BEFORE

Zoltan Paul Dienes

UPFRONT PUBLISHING
LEICESTERSHIRE

I WILL TELL YOU ALGEBRA STORIES
YOU'VE NEVER HEARD BEFORE

ISBN 1 84426 191 3

First published 2002 by
MINERVA PRESS

Second Edition 2003 by
UPFRONT PUBLISHING
Leicestershire

I WILL TELL YOU ALGEBRA STORIES YOU'VE NEVER HEARD BEFORE

I Will Tell You Algebra Stories You've Never Heard Before

This is a book about playing with powers,
Two, three or five alternating as base;
With logic blocks and strange words and some towers
You will be playing when problems you face.

You'll be constructing a great many cycles
Based on some rules that you too can invent;
Your strangest rules can have different titles,
They'll be the toys that will keep you content.

One long succession that I have suggested,
Takes you to feast in some tropical climes;
Luau for sixty-three people's requested,
Rules then determining how each one dines.

Pentagon Island for some shipwrecked sailors –
All of four races there harmony find;
Fish in the sea they will share like good whalers,
Building their houses to suit every kind.

With your strong arms you will point to the ceiling,
Or to the walls with them bent or quite straight.
Rolling some boxes with mirth you'll be reeling,
As painted stick men appear on each plate.

Learn mathematics through tricks of osmosis;
Knowledge creeps in through enjoyable fun.
Let these maths stories be your first oasis
In your maths desert when this book is done!

Preface

If you are one of the many who have sat through interminable boring algebra lessons, without being able to make head or tail of what was going on, but you may have had a slinking bit of curiosity as to what it all might be about, then the following pages are for you.

I have tried to introduce the reader to some of the mysteries of mathematical constructions, not through the senseless repetition of unintelligible jargon, but through the use of colours, shapes, words, sentences, bodily movements, music and, believe it or not, even through the use of numbers! To construct mathematics is the art of putting things together, while unravelling how the bricks in the construction relate to each other. The constructions will get more and more complex, more and more challenging, as you plough your way through these pages, so I suggest, in order not to get too frustrated, that you acquire or make the materials which you will need to physically manipulate in order to unravel the mysteries. Hopefully the mastering of each stage will make you more curious about what might be coming next.

The introduction is only intended for those of you who wish to know a little of the mathematical content of each section. If you are not interested in such information, you can omit the introduction and jump right into the first set of constructions! No actual previous mathematical knowledge is required to do so, apart from the simple arithmetic of the four operations. Further examples can be found on my website http://www.zoltandienes.com.

Contents

Introduction

The activities suggested in the following pages are based on constructions that mathematicians call groups, rings, fields and vector spaces. There is no need actually to know explicitly what these constructions are in order to carry out the activities; in fact, the reverse is more likely to be true: anyone who will have engaged in these activities is more likely to be able to understand these mathematical constructions.

I start with variations on the powers of the number 2, going up to the fifth and sixth powers. Then I come to the powers of 3 and 5, reaching the third power of these well-known little numbers. The last chapter is on the number 7, just so as to include all the one-digit prime numbers!

Chapters I and II

These activities include some fun with $2 \times 2 \times 2 = 2^3$ as well as with $2 \times 2 \times 2 \times 2 = 2^4$.

Let me explain briefly what I mean by $2 \times 2 \times 2$ or $2 \times 2 \times 2 \times 2$.

Take the verb 'to come'. Imagine it in the present or in the past; then think of it as in the first person or in the third person; then think of it either in the singular or in the plural. So for each of time, person and number we have two possibilities. This is where the number 2 comes in. Then we generate the third power of this number 2 by realising all the possible suggested ways of looking at the verb 'to come'. These are shown in Table 1.

Table 1

I come	He comes	I came	He came
We come	They come	We came	They came

So this is a $2 \times 2 \times 2$ or 2^3 situation, since there are two possibilities for each of three ways of looking at our verb.

Equally well, a set of red or blue circles or squares, big or small, would yield a 2^3 situation, since for each of the three properties colour, shape and size we have considered two possible values.

In order to create a 2^4 situation, all we need to do is to think of another way of looking at our verb or block. In the case of the verb we could think of direction as a fourth way of looking at the verb, two suitable values being *in* and *out*. Each of the eight possibilities in the 2^3 situation can have either an *in* or an *out* attached to it, thus creating sixteen ways of looking at our verb 'to come'. In the case of the blocks, we could have some thin and some thick, thus providing sixteen blocks.

There are systematic ways of putting our verbs in sequences. This is done by inventing a rule through which, by looking at any particular form of our verb, we can decide what the next form is going to be in the sequence.

The problem is also faced of comparing two different representations of our construction. This means, in the case of verbs and blocks, that there should be a reasonable rule through the use of which we can tell which block 'goes with' which verb form and vice versa. Such exercises are interesting as they pinpoint what is common to the two constructions which is independent of the particular representations used. This allows us to tease out what the rules of construction are, in other words the hidden mathematics that makes it that things happen as they do!

Chapter III

This entire section is devoted to 2^4 situations. The sixteen possibilities are provided as follows (where A stands for African, C for Chinese, E for English and I for Indian). The cells in Table 2 show sets of the four kinds of people.

Table 2

	A	C	AC	E	EA	EC	EAC
I	IA	IC	IAC	IE	IEA	IEC	IEAC

Many of the constructions in this section are to do with selecting parts of the above set of sixteen cells.

A triad is a set of three cells in which there is a total of an even number of each kind. If we add the empty cell, a triad has four members. For example ACE, CI, AEI is a triad. Also A, C, AC is a triad. (Do not forget that zero is an even number!)

A septet is a set of seven cells in which out of a certain set of letters there is a total of an even number of each letter of the set. If we add the empty cell, a septet has eight members. For example ACE, CI, AEI, EA, ACEI, C, I is a septet as A and E occur twice or zero times in each set. Also A, C, AC, E, EA, EC, EAC is a septet as I occurs zero times, zero being an even number.

A number of ways are explored of placing septets symmetrically, in five-way, two-way and three-way symmetries. Different ways of holding the right and the left arms are given as another way to embody a 2^4 situation. The left arms can be raised or level; in each of these positions it can be straight or bent. Using the right arm as well leads to sixteen ways of holding the arms. Cyclic sequences are also suggested, the succession rule being given away! Others can naturally be invented by the reader. At the end of the chapter some mathematical explanations are given for those who might be interested.

Chapter IV

This section is also devoted entirely to the 2^4 situation. It is suggested that the reader should make a set of cards, with pictures on one side and numbers on the other side. The activity can be played as a guessing game, looking at one side of the card and trying to guess what is on the reverse side. Of course if you read the given rules beforehand, the guessing game is no longer to do with guessing, because the given rule is used to deduce what is on the reverse side.

Rectangular arrangements are also suggested, with triads in the columns and the rows. The cyclic sequence of fifteen of the pictures is also extended to a cycle of arm positions.

Chapter V

It is not necessary to look at sixteen elements only as a concrete embodiment of the fourth power of the number 2. It could also be considered as an embodiment of the second power of the number 4. Of course, the number 4 itself is the second power of the number 2, so using mathematical jargon we can write $2^4 = (2^2)^2$.

For example, we could consider a dessert knife and a dessert fork, as well as a main dish knife and a main dish fork. There are two kinds of utensil: knife and fork. These are symbolised by the 2^2 in brackets. Then there are two kinds: dessert ones and main dish ones. These are symbolised by the second power outside the brackets.

There are sixteen ways of setting a place, including the place with all the utensils as well as the one with no utensils (this person can use his or her fingers). It is convenient to place the knives and forks in a cyclic order like this:

knife → fork → knife and fork → knife

I shall denote the dessert utensils with lower case letters, and the main dish utensils with upper case letters, so the objects we are considering are k, f, K, F.

Here is an example of the kind of rule that one can invent with the aid of which we can place the settings round a round table. The rule leaves out the empty setting.

(i) Rule for the next dessert utensils. Look at the present setting (choose any one of them). If there is just one knife there, there will be a dessert knife in the next setting. If there is just one fork there, there will be a dessert fork in the next setting.

(ii) Rule for the next main dish utensils. Look at the same present setting. Make the main dish setting next in the cycle and put it together with the present dessert setting. If there is just one knife in the above, there will be a main dish knife in the next setting. If there is just one fork in the above, there will be a main dish fork in the next setting.

For example, which setting comes after [k f K]?

By rule (i) there will be one dessert fork in the next setting. Rule (ii) instructs us to take the setting next to K, which is F (in the cycle) and put it together with k f, which makes the setting k f F. There is one knife in this set, so the main dish setting is K. So the setting next to [k f K] is [f K].

In this way the whole cycle can be worked out (not forgetting that 'next in the cycle' to 'nothing' is 'nothing'), which turns out to be this:

> [k f K], [f K], [k f], [k f K F], [F], [f K F], [k K F], [f], [f F], [K], [k F], [k f F], [k], [k K], [K F]

which you can check, if you like. It is just an example of the kind of rule that works. You can play with rules of your own and see what happens to the settings.

Of course, you may be wondering why we don't have spoons. You are right. We can have spoons as well, dessert ones as well as main dish ones. So the objects we have in this case are k, f, s, K, F, S.

We shall have to invent a cycle for these settings. We could have, for example:

$$k \rightarrow f \rightarrow k\,f\,s \rightarrow k\,f \rightarrow k\,s \rightarrow s \rightarrow f\,s \rightarrow k$$

(You might have noticed that in any run of four settings there are two or none of each kind of utensil if you miss out the third one in the run.)

We can now organise a luau for 63 people round an enormous table, each participant having a different setting!

You could try a rule something like this:

(i) Put together the present dessert utensils with those next in the cycle of the main dish utensils. Any utensil of which there is just one in this joint set will form part of the dessert utensils in the next setting.

(ii) Put together the present dessert utensils with those next but one in the cycle of the main dish utensils. Any utensil of which there is just one in this joint set, will form part of the main dish utensils in the next setting.

Take for example the setting [k f S]. By rule (i) we must put together [k f] with what comes after S in our cycle, which is [F S]. Putting these together we get [k f F S]. We now have to take those utensils of which there is only one in this set. These are the knife and the spoon. So by rule (i) there will be a dessert knife and a dessert spoon in the next setting round the luau table.

By rule (ii) we have to put together [k f] with the setting that comes in the next setting but one after S. This is a knife. So we have [k f K]. Only the fork occurs just once in this combined set, so that by rule (ii) the main dish setting is just one fork. So the setting we are looking for is [k f F]. If I am not mistaken, you will get all the settings for the luau in the sequence generated by the suggested rules!

Now you should play with rules invented by yourself and see what happens to the settings. In the knife–fork–spoon settings we have been immersed in a $(2^3) \times (2^3) = (2^3)^2$ situation! At the end of the section it becomes a 2^6 situation!

Chapter VI

In this chapter I deviate from the straight path of playing about with higher and higher powers of the number 2. There are no cyclic sequences, but creatures such as cubic polynomials and their factors make their sudden appearance, followed by a change of rules, resulting in even stranger creatures known as the quaternions. A particular case of these creatures is complex numbers, incorporating the infamous square roots of –1.

The reader might like to evolve a generalisation of the cubic polynomial constructions to polynomials incorporating higher powers by simply lengthening the cycles. Every polynomial can only be constructed in that way by the cycle becoming infinite, in fact by turning the cycle into an infinite sequence such as the one of the natural numbers.

Chapters VII and VIII

In these sections I exploit the fifth power of the number 2. The activities in these sections help us to realise the feel of a $2 \times 2 \times 2 \times 2 \times 2 = 2^5$ situation.

One way to create such a situation is through the use of attribute blocks (logic blocks), using the four shapes of triangle, rectangle, circle and square, big and small, thick and thin, red and blue (or any two of three colours). The shapes can be divided into two rectangular ones (rectangle and square), and two non-rectangular ones (triangle and circle). Another way of dividing them into two kinds is to look at the symmetries possessed by them. If we call anything greater than the number 3 a large amount, then we can say that the circle and the square have a large number of symmetries, whereas the triangle and the rectangle have a small number of symmetries. So there are rectangular and non-rectangular blocks; each kind can be either a high symmetry or a low symmetry block. In this way we have succeeded in turning our 4 blocks into a $2 \times 2 = 2^2$ situation. So our 32 blocks embody a 2^5 situation.

In this section I begin the activities with a sequence of 31 blocks, having taken out the zero block, which of course could be any of the 32 blocks.

Each of the properties symmetry, rectangularity, size, thickness and colour has two values. One of these is chosen to be a strong value, the other one being the weak value of that property. The block all of whose properties are weak has been chosen to be the zero block. This is the small thin blue triangle. So this block is left out of the sequence of the 31 remaining blocks.

As you go round the cycle, it will be noticed that the way strong and weak properties alternate is exactly the same for all five properties. There are exactly two such possible rules of alternation between strong and weak. In the case of a 2^4 or in the case of a 2^3 situation, there is only one such rule of alternation. If you happen to be a mathematician, you might try to prove this assertion!

The search for subsystems gets a little more challenging here. There are 31 subsystems of fifteen blocks each, which can all be placed in cycles of fifteen obeying the alternation rules of 2^4 systems. It is not without interest to look at the intersections of any two of these 15-systems. Perhaps not surprisingly the intersection is always a 7-system! What we are doing is investigating the subspaces of a five-dimensional space. The 15-systems are four-dimensional, the 7-systems are three-dimensional and the triads are two-dimensional. A one-dimensional space would consist of one block. To each of these subsystems you would really have to add the zero block, so that each subspace would really have one more block then before! These numbers of blocks are, of course, the respective powers of the number 2.

In chapter VIII the activities are essentially repeats of chapter VII, but using 32 sentences instead of 32 blocks. Then the sentences are replaced by little towers of patterned cubes, using naturally five different patterns. Subsystems of fifteen are looked at in a symmetrical arrangement of the towers, using five lines of symmetry.

Chapters IX and X

I now deviate from the straight and narrow of the number 2 and smuggle in the number 3, what is more, to the third power. But out of the 27 elements obtained as 3^3, I shall only need 13. But which 13? This problem is tackled in the first part of this chapter. Then I select 13 little towers out of 27 possible ones, and finally I propose 13 numbers, which behave in just the same way as the selected blocks or the towers.

The number 13 will eventually be used to tease out the 13 axes of rotation of the cube. There are subsystems in these activities which will be identified with those axes of rotation which all lie in the same plane – well, not all of them, but I won't give the game away just yet. Read the chapter and you will get some surprises.

In chapter X I lead the reader into finite geometries. You will have guessed it! Our geometry will have exactly 13 points, as well as exactly 13 lines. Each point will belong to four lines and each line will have four points on it. The question is left until later whether other numbers besides the 13 can be used for constructing finite geometries. I examine Desargues's theorem on perspective triangles as well as the construction of the cross-axis of two ranges of points. All the above is done through arranging ways in which 13 players can play tennis, but so that every player gets on the court with every other player, but only once. Naturally, you will need 13 tennis parties!

Chapters XI and XII

Here I go back to the number 2, temporarily abandoned in the last two chapters. The situations explored here come under the heading $(2 \times 2) \times (2 \times 2) \times (2 \times 2) = (2^2)^3$.

Imagine groups of boys and girls, men and women and old men and old women. In each age group you could have either nobody, or just one male person, or just one female person, or just one of each. So there are 64 ways of making such groups. If we do not count the group with nobody in it, then there are 63 possible groups of people, made up of component persons as described above.

Three groups will get together to play a game such as Scrabble against each other. They will sit round a triangular table, each group being spread out along one of the sides of the triangle. The three groups must be so chosen that if you go round the triangle in the clockwise sense, for each age group the presence of just one male is followed by the presence of just one girl, which is followed by the presence of one of each sex, after which we return to the one boy in that age group. In some games there will be only one age group, in others there will be two, and yet in others all three age groups will be present. Since there are 63 groups, this means that 21 games will be played.

We can add two groups by simply putting them together, but removing any two persons of the same sex and of the same age group. So the sum will contain only the kind of persons that are found in either one or the other of the groups to be added but not in both.

We can also multiply a group by a male, a female, or a couple. Multiplying by a male will leave the group unaltered. Multiplying by a female will replace each male by a female, each female by a couple and each couple by a male, always of the same age group. Multiplying by a couple will replace each male by a couple, each couple by a female and each female by a male, always of the same age group. Multiplying by the nobody will yield the empty group with nobody in it.

You should now check the following:

(i) Adding any two of the three groups playing a game, the sum will always be the third group.
(ii) Multiplying one of the groups playing a game by female, you get the group next to it by going round to it in the clockwise sense.
(iii) Multiplying one of the groups playing a game by a couple, you get the group next to it by going round to it in the counterclockwise sense.

We can now go for the subsystems. Adding two groups from two different tables, you will get a group on a third table. Now add any two groups from the three tables you now have. You may find it on one of the three, or you might be able to add another, fourth, table. If you go on like this, you will eventually collect five tables which are such that any two groups among them will add up to a group on one of the five tables. So the five tables obtained in this way will form a subsystem. There will also be a property which all the groups in the subsystem have, which none of them have that are outside the five tables.

Some fun can be had in organising these sets of five tables in various ways, such as embedded in a square or hexagonal shape of tables. The activities are not actually given with the aid of games tables, but with the aid of discs with pictures of starfish and coconut trees. In chapter XII the disks are replaced by prisms. Rules are suggested for constructing sequences of these elements, leading up to the possibility of making a sequence with 63 of the groups!

A brief set of mathematical notes for the curious ends chapter XII.

Chapter XIII

Here we return to Desargues's theorem and the construction of the cross-axis in a finite geometry of 21 points.

An algebraic-looking coding system is suggested for the players, which relies on the properties of a $(2^2)^3$ situation explored in the previous two chapters. Duality is introduced through appointing umpires or scorers for the Scrabble teams, and another coding system is suggested for identifying each group of Scrabble players.

This chapter is heavy reading, but great fun when it is mastered. You should have the activities of the previous two chapters at your fingertips before embarking on this version of Desargues's theorem!

Chapters XIV and XV

This is where our variations on a theme of number 2 come to an end. We are now leaping forward to see what can be done with the number 5, and what is more, we shall start right off with 5 to the third power! So in these two chapters the activities create 5^3 situations.

This is at first expressed through the use of five people, who now have the boring algebraic-sounding names of {0, 1, A, B, C}, which you can replace by Zebedee, Wang, Albert, Betty and Charlie, if you prefer! But the names are soon replaced with patterns – visually much more acceptable!

Adding is again defined but in a somewhat weird way, as well as multiplying. Non-mathematical readers might well wonder why on earth I would have thought of these ways, which look like *deus ex machina*! But if you patiently plough through my abstruse-sounding suggestions, you will see that some very interesting things come to light in the end!

In chapter XV very similar feats are attempted, but through the use of arm positions. You will be asked to make boxes, on each of four sides of which there will be three persons in various arm positions.

Since there are three persons on each side of each box, and each person can assume five different positions, so there are 125 ways that we can make a set of three persons. The one in which all have their arms stretched out horizontally is left out, this being the zero position, thus leaving 124 pictures. Since there are 4 of these on each box, there will be 31 boxes. As you roll the box away from you, the arm positions of the persons will change following the same rule of succession. This means that once you have decided on how to draw the first picture on a box you are making, the rule will tell you precisely how to draw the other pictures on the remaining three faces of the box.

How the activities with the boxes are just like the activities with the five persons or the five patterns is for you to find out. If you have got as far as this set of activities, your brain should be twisted enough to enable you to untwist such problems!

Chapter XVI and XVII

Now we do a small leap backwards and start to play about with the number 3. In chapter 16 a 3^2 situation is explored at depth, while in chapter 17 we graduate onto a 3^3 situation.

A 3^2 situation clearly contains nine elements. These can be expressed using the numbers 0, 1 and 2. For example we could imagine these little towers, symbolised in the table on the left, the first digit denoting blue squares, the second digit red squares.

21	01	11
20	00	10
22	02	12

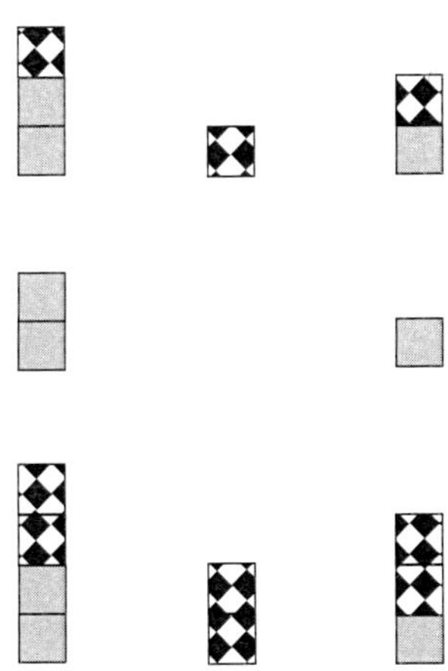

Rules are developed for passing from one tower to another tower. In the application of any rule, if a number greater than 2 is obtained, we shall have to subtract 3 or some multiple of 3 from the result, as otherwise we would not find any tower corresponding to our result. Let us use B to indicate the number of blue squares and R the number of red squares in a tower.

Then one rule for getting to another tower could be:

B next = 2 × B + R
R next = 2 × B + 2 × R

which would send us careering among the towers as shown below.

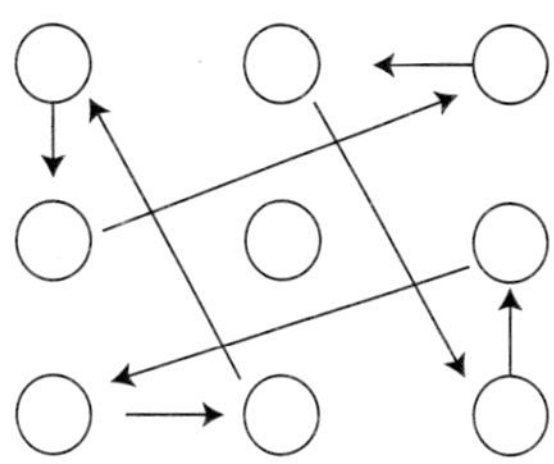

There are 48 such ways of deciding where to go from any given tower, which never lead to the zero tower in the middle.

A guessing game can be played by picking a rule and then finding the path it tells us to follow on the map of the towers. Or you can pick a map and try and guess the rule that will make us perform those very moves shown on the map. It is possible to identify a number of interesting subsystems within the set of 48 rules or functions. There is a set of 8 rules, which relate to each other exactly in the same way as do the 8 quaternions discovered earlier. You can even find subsystems that behave just like the rotations and symmetries of the equilateral triangle, of the square and of the regular hexagon. Unfortunately there does not exist a set of 12 rules which behave like the rotations of the regular tetrahedron.

In chapter XVII we extend our horizons to a 3^3 situation. This is done by drawing pictures of daisies, tulips and sunflowers, never more than two of any kind of flower on any one picture. If we leave out the zero picture with nothing on it, we are left with 26 pictures. These can easily be put into pairs. A pair of two pictures must have a total of either zero or three of each kind of flower, counting both pictures together. This means that we now have thirteen pairs, and we can now see the origin of our previous thirteen game!

The 13-element finite geometry is developed again, the points being the pairs of pictures, the lines being the sets of four pairs that belong together in virtue of some rule or rules. Names are suggested for these sets of four pairs, reminiscent of the reason why they are supposed to belong together. Sequences taking in all 26 pictures are suggested, as well as sequences of pairs.

Chapter XVIII

We finally reach the number 7! I could not resist trying to make a connection between seven ways of holding our arms and the seven notes in the major scale. Adding and multiplying are introduced, with the multiplication table given on the last page showing how one arm position can be multiplied by another arm position.

Ways are also suggested of writing a second voice to a tune. The method suggested becomes particularly successful for tunes in the pentatonic scales.

Chapter I

SEVEN PICTURES GAMES

1. Introducing the Pictures

There are some interesting things we can do by taking three objects and making all the different kinds of sets with them. Let me take the following three objects (see Figure 1).

Figure 1

It is possible to make the following sets with our three elements, not counting the set with nothing in it (the so-called empty set) (see Figure 2).

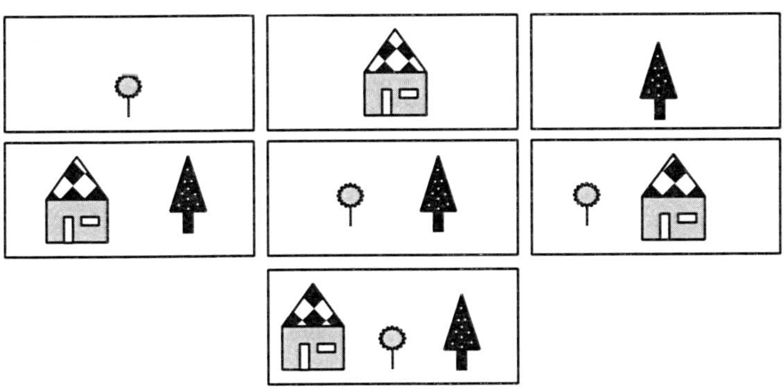

Figure 2

I shall call each of these sets of objects a picture. These pictures will be our toys with which we shall play some games, or solve some problems, if you prefer to sound more serious!

2. Adding Pictures

The addition of two pictures is defined as follows:

The sum picture of any two pictures A and B consists of those elements which are present either in picture A or in picture B but not in both (see, for example, Figure 3).

Figure 3

It is worth noting that out of the above three pictures any two will add up to the third one. You will find that this is so in any addition done by our rules. It means that we are in some way enclosed within our set of three pictures. There is no way we can get out of this set by adding any of the pictures to each other. We need to choose a third picture and then, by successive addings, we can reach all the other pictures.

A set of three pictures such that the sum of any two of them is equal to the third one will be called a triad. There are seven triads that it is possible to make out of our seven pictures. Any two triads will have a picture in common. Any picture belongs to exactly three triads. To see this more clearly, we can place the pictures in a triangular pattern such as the one shown in Figure 4.

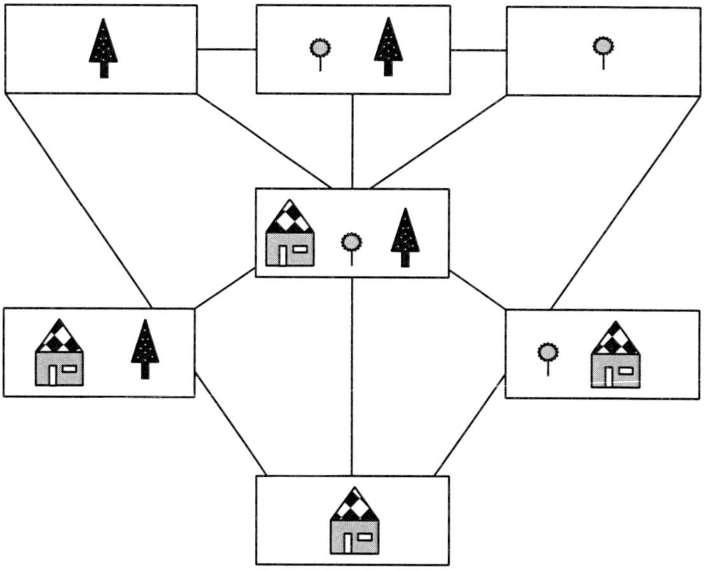

Figure 4

Six of the seven triads are situated along straight lines. The seventh one is in the middle. It is easy to see the three triads to which each picture belongs (not forgetting the triad in the middle).

There are several other ways of arranging our seven pictures. For example, try to do the following:

(i) Take one triad away from your set of pictures.

(ii) Call the pictures in your triad A, B and C.
(iii) You will be left with four pictures: D, E, F, G.
(iv) Arrange these four pictures in the form of a square:

D E
F G

(v) Exchange the positions of the pictures making up your square until the following are all triads:

[D E A], [F G A], [D F B], [E G B], [D G C], [F E C]

Another way of arranging the pictures would be in the form of a circle. As you go round the circle you should meet every element three times in succession; you should then come to a picture without this element, after which there should be a fourth picture with this element in it. So each one of the elements should be in the picture or not, according to the rhythm:

Yes Yes Yes No Yes No No

where Yes means that the element in question is in the picture, and No means that the element in question is not in the picture. A solution to the problem can be found in Figure 5.

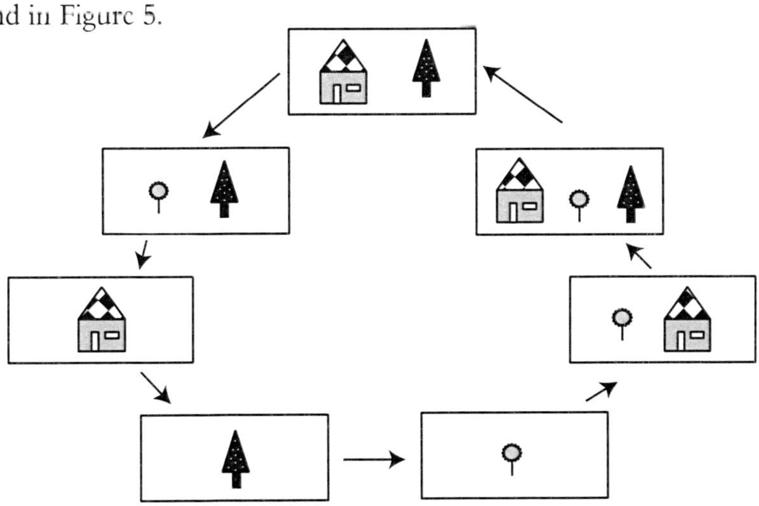

Figure 5

It is of interest to note that there is a rule that tells us which picture comes after any given picture. To work out the next picture, we only need to know what precisely is in the picture we are looking at.

Here are the rules:

(i) If there is a house in a picture, then there is a tree in the next picture.

(ii) If there is a flower in a picture, then there is a house in the next picture.
(iii) If there is a flower or a tree in a picture but not both, then there is a flower in the next picture.

Using the above rules, it is possible to cover up all the pictures except one, and work out what must be in the next picture without looking at it.

Every time we make a circle of our seven pictures with the given Yes–No rules, there will be some rule which will allow us to work out the picture after any given picture. It will not necessarily be the above set of rules; you might be less lucky and the rules might be more complicated.

You can also play the game with words, or with logic blocks. Here are some words you could use:

bin, fan, fin, ball, bill, fall, fill

Instead of house, tree and flower, you could use the letter *f*, the letter *i* and the ending *ll*.

Or you could use attribute blocks, using two shapes, two sizes and two colours. You would take out one of the blocks, and you could use the remaining seven blocks for doing the things described above.

You could even set up some more interesting links between these different ways of playing the games. Here are some examples:

bin fan bill fin fill fall ball

The above simple objects lead to four questions:

(i) What is the next word to any given word?
(ii) What is the next block to any given block?
(iii) Which block is below a given word?
(iv) Which word is above a given block?

You can ponder over these questions, but if you do not feel you want to exert your brain too much, here are the answers to the questions:

(i) After a word with *i* comes a word with *f*; after a word with *f* comes a word ending in *ll*; after either *b–ll* or *f–n* comes a word with *i*.
(ii) After a dark circle or a light square comes a square, but after a light circle or a dark square comes a circle. After a big circle or a small square comes a dark one, but after a small circle or a big square comes a light one. If a block is different in one or three ways from a small light circle,

then a big block comes next, but if it is different in two ways from a small light circle, then a small block comes next.

(iii) Below a word with *f–n* or *b–ll* you get a square, but below *f–l* or *b–n* you get a circle. Below a word with *f* you get a big block, but below a word with *b* you get a small block. Below *–in* or *–all* you get a dark block, while below *–an* or *–ill* you get a light block.

(iv) Above a big block you get an *f* word, but above a small block a *b* word. Above a block with one or three differences from a small light circle you get a word with *i* in it, but if it has two differences, you get a word with *a* in it. Above a big circle or a small square you get a word ending in *–ll*, and above a small circle or a big square you get a word ending in *n*.

Every time you arrange the words and/or the blocks to follow the given Yes–No rule, you will get some succession rules and linking rules such as the above.

The same is naturally true of the pictures. You could arrange the pictures alongside the blocks in circular patterns next to each other, and you would obtain similar results. One way is illustrated in Figure 6.

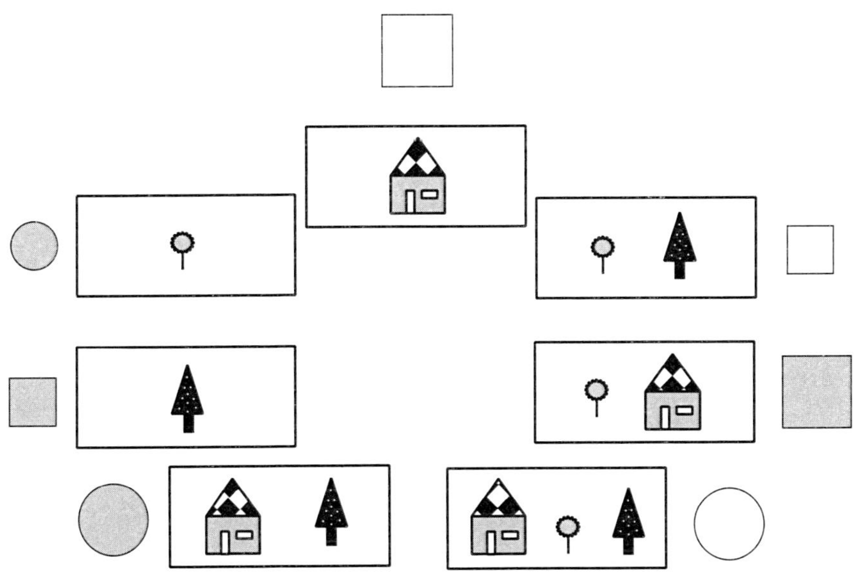

Figure 6

Chapter II

WHAT HAVE SENTENCES, NUMBERS AND BLOCKS GOT IN COMMON?

1. The Three Sentences Problem

Look at these eight sentences:

> I am going. I was going. You are going. You were going.
> I am coming. I was coming. You are coming. You were coming.

Four of them have the verb 'to go', the other four have the verb 'to come'. Four have 'I', the other four have 'You'. Four are in the present; the other four are in the past.

Write down three of the above sentences but in such a way that either just one of them or all three of them are in the present, also so that either just one of them or all three of them have 'I', and also so that either just one of them or all three of them have the verb 'to go'. Do not use the sentence 'I am going'.

Such a set of three sentences is called a triad. Try to write seven different triads. Any two of your triads will have a sentence in common, and any two sentences will belong to just one of the triads. If that is not so in your case, you must have made a mistake somewhere. Try to find it and correct it.

Write your sentences on pieces of paper. Put 'I am going' in front of you and place the other seven sentences in a circle round 'I am going', but make sure that in any run of four sentences round your circle, the first one, the second one and the fourth one form a triad.

Between any two sentences there might be one difference, or two differences or three differences. These are tense difference (present vs. past), verb difference (whether 'go' or 'come') and person difference (whether 'I' or 'You').

Look at one of your triads. Find what differences there are between 'I am going' and one of the sentences in your triad. Then find the differences between the remaining two sentences in your triad. Are these the same differences? Sometimes? Always?

2. The Three Numbers Problem and its Link with the Three Sentence Problem

Now look at the numbers 0, 1, 2, 3, 4, 5, 6, 7.

Let us call 0, 1, 2 and 3 small numbers and 4, 5, 6 and 7 big numbers. The numbers 0, 2, 4 and 6 are even, but 1, 3, 5 and 7 are odd. The names of the numbers 2, 3, 6 and 7 begin with either t or s, so we can call them S–T numbers, whereas 0, 1, 4 and 5 are Z–O–F numbers

Now let us make some triads out of our numbers. In a triad there must be one or three small numbers, one or three even numbers and one or three Z–O–F numbers. Number your sentences from 1 to 7, calling 'I am going' the 0 sentence. But number them in such a way that the numbers of the sentences in a sentence triad always form a number triad.

Then try to find a rule which will tell you the number of a sentence, and another rule which will tell you the sentence of a number.

Here is one solution. Don't look, if you want to solve the problem by yourself!

Here is a sequence of numbers in which, in any run of four, the first, the second and the fourth numbers form a triad. When you get to the end of the row, you start again at the beginning!

> 1, 2, 4, 3, 6, 7, 5

A sequence of sentences could be, for example:

> I was going. You were going. I was coming. You are going. You are coming. You were coming. I am coming.

It is easy to see that to every big number there corresponds a sentence with the verb 'to come' in it, and to every small number a sentence with the verb 'to go' in it. To every S–T number corresponds a sentence with 'You' in it, and to a Z–O–F number a sentence with 'I' in it.

If a number is odd or an S–T or big or all three of these (but not just two of them), then the sentence will be in the past tense. If a sentence is different in one or in three ways from 'I am going', then its number is odd. If the sentence is different from 'I am going' in two ways, then its number is even.

To a 'You' sentence there corresponds an S–T number; to an 'I' sentence a Z–O–F number. To a 'to come' sentence corresponds a big number, and to a 'to go' sentence a small number.

If you change the sequence by putting the last number first, there will still be a rule which will allow you to pass from number to sentence and from sentence to number. You could try, for example:

> 7, 5, 1, 2, 4, 3, 6

and find the rules connecting the sentences and the numbers.

3. Our Numbers and the Factors of 42

You can do similar things with the factors of 42. These are:

1, 2, 3, 6, 7, 14, 21, 42

Each of these numbers may or may not have 2 as a factor, may or may not have 3 as a factor, and may or may not have 7 as a factor. The numbers 2, 3 and 7 are the prime numbers on which the sequence is based.

Any set of three of the above numbers not containing 1 will be called a triad if each of the three prime numbers occurs either not at all or exactly twice in the set. For example: {2, 7, 14} is a triad since 3 is missing altogether, 2 occurs in 2 and in 14 but not in 7, 7 occurs in 7 and in 14 but not in 2.

Now make a sequence of these numbers (excluding the number 1) so that if you put them round in a circle, in any run of four numbers following one another, the first, the second and the fourth numbers always form a triad.

Then write your sequence of numbers from 1 to 7 alongside your sequence of the factors of 42, so that these factors will be numbered 1, 2, 3, 4, 5, 6 and 7. Find the rule for telling the number of a factor, and for telling the factor of a number. Here is one solution. Don't look at it, if you want to solve the problem yourself. There are lots of different ways of solving the problem – I am only giving you one of them!

2, 3, 42, 6, 14 ,7, 21
1, 2, 5, 3, 7, 6, 4

You might have noticed that, by chance, to every even number in the first row there corresponds an odd number in the second row, and to every odd number in the first row an even number in the second row! It is not hard to see that to every 7 number in the first row there corresponds a big number in the second row, and to every non-7 number in the first row a small number in the second row. Finally, any number in the first row which has just one of the factors 3 and 7 stands above an S–T number, while any number in the first row which either has both or neither of these two factors, stands above a Z–O–F number. If you want to make sure that you will find a 3 number in the first row, choose either a small S–T number or a big Z–O–F number from the second row.

Look at the first row again. There are three even numbers following one another, separated on one side by one odd number and on the other side by two odd numbers from the fourth even number. If you look for the 3 numbers, you will see these are also bunched in the same way. By a miracle, so are the 7 numbers! In each case you have a cycle that can be described as yes, yes, yes, no, no, yes, no.

Now try to put your 7 numbers into another cycle, making sure that the 2 numbers (meaning the even numbers), the 3 numbers and the 7 numbers follow the cycle Yes Yes Yes No No Yes No. You will find that in any run of 4 numbers in your cycle the first, the second and the fourth numbers will form a triad.

4. The Three Blocks Problem

Now you can invent another game, using the following attribute blocks:

Small yellow circle   big yellow circle   small red circle   big red circle

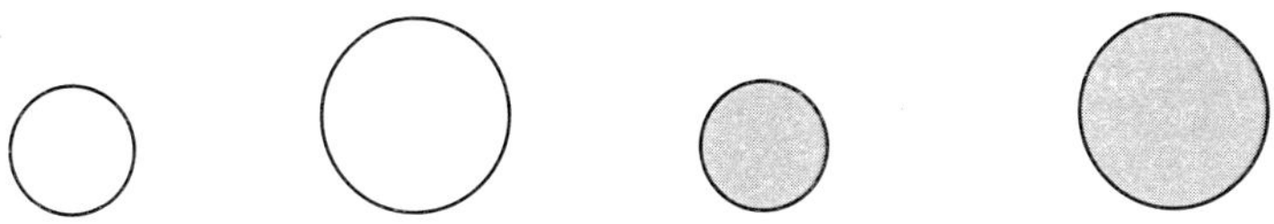

Small yellow square   big yellow square   small red square   big red square

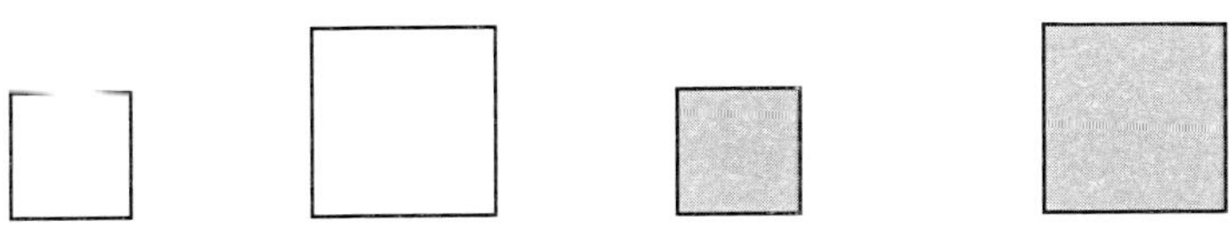

Of course, you will have to take out one of the blocks, which will then correspond to 'I am going' in the sentence game, to 0 in the 0, 1, 2, 3, 4, 5, 6, 7 game, and to the number 1 in the game with the factors of 42. You can do the same things with these blocks that we have already done with the sentences and with the numbers. For example, leaving out the small yellow circle, we can construct cycles such as:

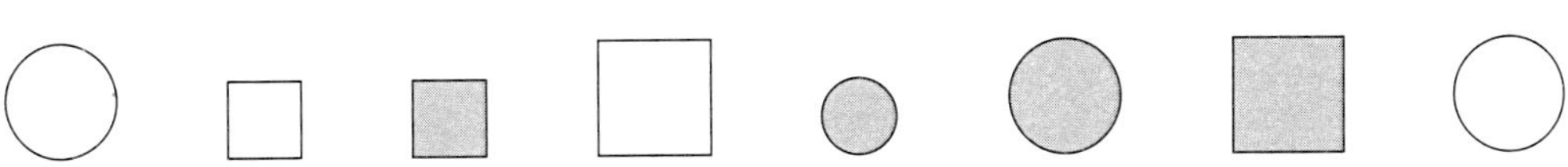

Try to relate these blocks to the numbers 1, 2, 7, 3, 5, 4, 6, 1 in that order. For example, a big yellow or a small red block will always go with an odd number, and the red ones always seem to go with big numbers!

5. Longer Cycles

You might have been wondering whether you can make longer cycles using more sentences, more numbers or more attribute blocks. The answer is, of course, in the affirmative. For example, you can use the factors of the number 210. These are:

1, 2, 3, 5, 6, 7, 10, 14, 15, 21, 30, 35, 42, 70, 105, 210

In this case a triad is a set of three numbers such that each of the factors 2, 3, 5, 7 occurs in just two of the three numbers or is missing entirely. For example {6, 10, 15} is a triad, since 7 is missing, 2 occurs in 6 and 10, 3 occurs in 6 and 15 and 5 occurs in 10 and 15.

First problem: make five triads, but such that no two of the five triads have a common number.

Second problem: put your five triads in five columns, thus making a 3 × 5 rectangle of numbers. Then arrange the order in which the columns are written in such a way that the first three numbers in any row, as well as the last three numbers in any row, form triads. You should end up with your original five-column triads together with six more row triads.

Third problem: choose any two of your five columns. Call them A and B. Choose any number but not out of columns A or B. Call it C. Find a number in the triad A, and a number in the triad B such that by joining the number C to them you will have made a triad. However you choose C (outside A and B), you will always be able to find an A-number, and a B-number, which together with C will make a triad.

You could make sixteen sentences out of our set of eight sentences by putting 'home' or 'away' at the end. In this way we place a complement to the verb, so we have the possibility of making a complement difference. You could, for example, appoint the sentence 'I am going home' as the key sentence. So in a triad of sentences you would have to have one or three 'home' sentences, one or three 'I' sentences, one or three 'present' sentences, and one or three 'to go' sentences.

You could set the same problems as before, but with our new definition of a triad.

Fourth problem: place the sentence 'I am going home', together with the number 1, on the table in front of you. Then put the remaining fifteen sentences and numbers round the key sentence, but in such a way that in every run of five sentences and/or of numbers, the first, the second and the fifth ones always form a *triad*.

When you have done this, try to find a rule that tells you which number you have placed with which sentence! Table 3 hows a solution for the two cycles and the rule for finding the number that goes with any particular sentence. Don't look if you want to solve the problem yourself!

Table 3

I am going away	You are going home	I was going home	
21	30	105	
I am coming away			I am coming home
10			210
I was coming away			You are going away
42			70
You were coming away			You were going home
35			14
You were coming home			I was coming home
15			2
You were going away	You are coming home	I was going away	You are coming away
6	7	5	3

6. The Relationships Between the Sentences and the Numbers

I coming or You going → factor 2

One or three differences from key sentence → factor 3

One or three differences from key sentence not counting complement difference → factor 5
One or three differences from key sentence not counting person difference → factor 7

Table 4 shows a solution to the 3 × 5 rectangle problem.

Table 4

14	42	3	7	21
105	70	6	35	210
30	15	2	5	10

You can play the game of choosing two columns, say the third and the fourth columns. Any number outside these two columns should be obtainable as the third number of a triad, one of which is in the third column, the other being in the fourth column.

Chapter III
THE SHIPWRECK PROBLEMS

Some unlucky mariners get shipwrecked on an island. There are eight Africans, eight Chinese, eight English and eight Indians. They build fifteen houses for themselves, as shown by the circles on the map below (not counting the circle in the centre of the island). Since there is no chance of rescue, they decide that they have to learn to get on with each other; they settle in the houses in the following way:

The leader of each group will live by himself in a house. In the other houses, either two mariners move in, or three, and in one house four mariners move in. No house must have two of the same race living in it and no two houses must have the same combination of races in it (see Figure 7).

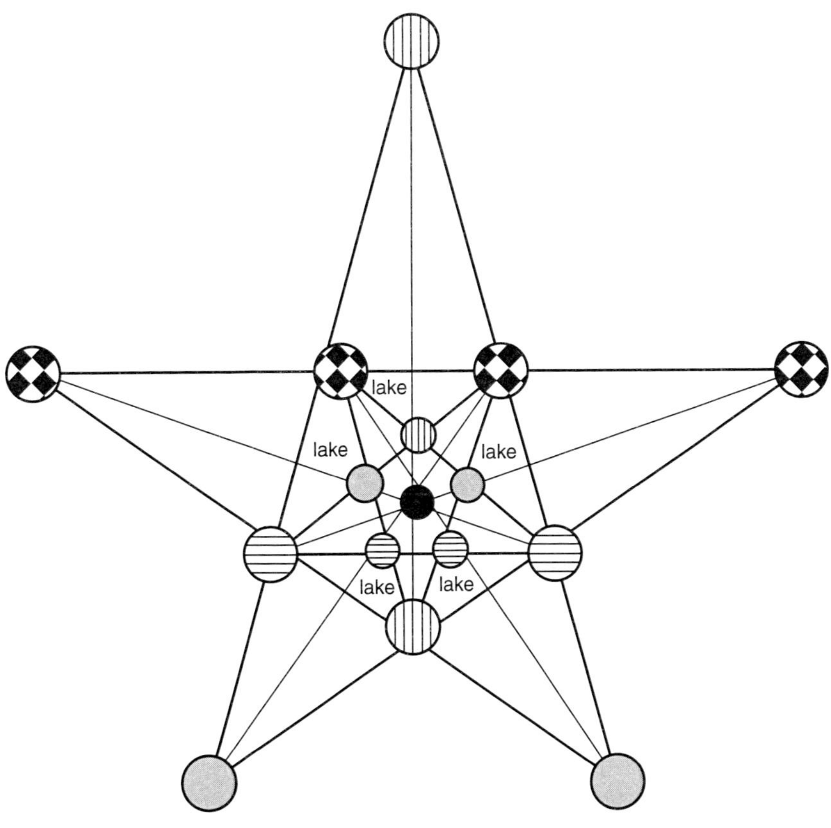

Figure 7

The shipwrecked people have other problems before they can settle in. The five bays between the five sharp points abound in very good fish. Round each of these bays there are three houses, one at the inner part of the bay and two on the points on either side. To avoid arguments, they decide that the total number of persons of any race living round any of the bays must be either two or zero.

There is also some very good fish in the five lakes marked on the map. They decide that the same rule must apply to the three houses built on each of the five lakes.

To avoid problems of communication, they construct five roads right across the island (using rafts to cross the lakes). Again to avoid problems they decide that along any of the five roads there must be two or none of each race in the three houses along each of the five roads.

With these rules they manage to live very happily, as the island is very fertile, apart from the abundance of fish.

Copy the map on a large piece of paper, and use coloured or patterned cubes to represent the various races, making a tower of the cubes for each house. Can you arrange them according to the rules they invented?

The 'two or none of each race in three houses' eventually became a way of getting together for a party. They built a clubhouse right in the centre of the island (shown black on the map), and they decided that in order to have a party (unless everyone from all the houses came), they would have to have people from three houses which satisfied the 'two or none' rule. Such a set of three houses they called a triad.

You already know some triads: five round the bays, five round the lakes and five along the roads. There are lots more! Try to find them! You should be able to locate twenty more triads. Find where they are on the map!

At one point the islanders decided to have larger parties, but still not so large as to involve all the inhabitants. They found a way of involving seven houses. The seven houses had to be so chosen that the set of houses should be made up only of triads. This meant that if the people from any two of the houses invited wanted to talk to the people of the house which was the third house of the triad, the people from that third house were in fact at the party.

Here is an example. Suppose you wanted to invite the people from the houses ACI, A, ACE then we would have to complete all possible triads starting with any two of the above. The first two would be completed by CI, the last two by CE, the first and the last by EI and ACI and CE would give AEI. So we would have the seven houses (see Figure 8).

ACI	A	ACE	CI	CE	EI	AEI

Figure 8

The house that completes any two of these seven houses so as to become a triad is already there among the seven houses.

The islanders called such a set of seven houses a septet. On the map, the houses with vertical stripes, together with the houses of any one of the other patterns, can form a septet.

It is perhaps worth noting that in the above septet, in every house there are either two non-Africans or zero non-Africans (the latter being the case for the second house in the row).

In fact, every septet can be identified by such a property. So when the septet party invitations went out, it would be enough to say, for example, that people from those houses were invited in where either there was an English person as well as an Indian, or there was neither an English person nor an Indian.

You will find that there are exactly seven houses fitting the above description. Try to find them! Make sure that this set of seven houses consists entirely of triads. What about the set of all houses in which there is an even number of inhabitants? Will they form a septet? Find out!

There are in total fifteen septets. Try to find them. Find the property of each septet which distinguishes its houses from all the other houses on the island. How many houses do two septets have in common? Is it three? If so, do they form a triad? What about three septets?

Here is the solution to the property problem. A seven house party can have a property in one of the following ways:

(i) Members of one race have not been invited.
(ii) In every house of the seven, of two given races either both or neither are resident in the house.
(iii) In every house of the seven, of three given races, either two or none are resident in the house.
(iv) In every house of the seven there is an even number of residents.

You can check whether you have a legal party by counting the people from each race invited to the party. If it is not four or zero, you do not have a legal party!

There is another interesting way of checking whether you are legal. Try to arrange the houses in a circular pattern so that out of any run of four consecutive houses, the first, the second and the fourth form a triad. If you can do this, you have a legal party. If not, your party is not legal. An example is given in

Figure 9.

Notice that all the patterns occur following the cyclic order:

is, is, is, isn't, isn't, is, isn't

You could make the sequence by requiring that rule for all the patterns.

This is the septet party in which in every participating house there is a Chinese and an Indian resident or neither.

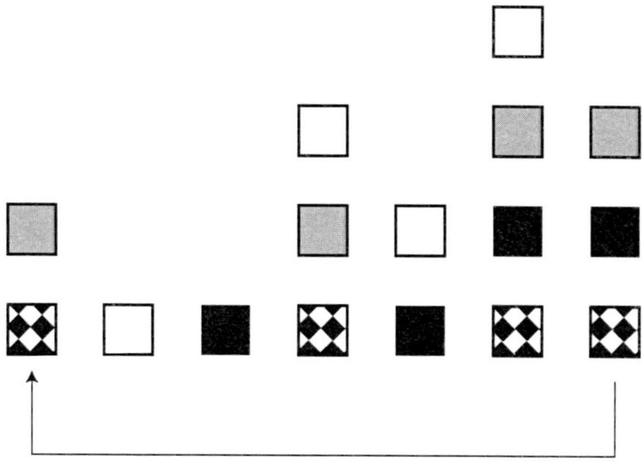

Figure 9

You could practise making septets by completing triads starting with three houses that do not form a triad and then arranging the houses in order. You could then find the property that distinguishes this septet from all other houses on the island. Or you could start with putting seven houses in order so that in every run of four the first, the second and the fourth houses should form a triad, and then find the property of the septet so formed.

Here is the solution to the 'find all the triads' problem. We shall have to invent some names for the various parts of the island. Let us say that if we join the five extreme points by drawing lines across the sea (thus closing the bays), we obtain a big pentagon. The five-pointed star joining the extreme points is the big star; its points will be called star points. The inside part of the big star is another pentagon, so let us call this the middle pentagon. The vertices of this pentagon are the inner points of the bays. The diagonals of the middle pentagon allow the formation of the lakes and they form the small star, inside of which we find the inner pentagon. The houses round this inner pentagon stand at the inner points of the lakes.

Having established the ways in which we can refer to parts of the island, you can now check that all the following are triads:

(i) Thin triangle: a side of the inner pentagon together with the star point opposite.

(ii) Small triangle: a diagonal of the inner pentagon together with the inner point of the bay opposite.

(iii) Lake: a side of the middle pentagon together with the inner point of the lake.

(iv) Big triangle: a diagonal of the middle pentagon together with the star point opposite.

(v) Bay: a side of the big pentagon together with the bay point of the same bay.

(vi) Flat triangle: a diagonal of the big pentagon together with the inner point of the lake by it.

(vii) Road: a line of symmetry of the island.

Since there are 5 of each of the above kinds of triad, we can be sure that we have found 35 triads.

One day the clubhouse was struck by lightning and burned down. Then they decided that perhaps they did not need a clubhouse any more, since they could have the parties in their houses. So they made up a rule (in order to be fair) as to which septet should party in which house. Before you look at the way in which the islanders solved that problem, perhaps you could invent a solution of your own (see Table 5).

Table 5

Septet	House for party
Members of one race not invited	Chief of that race lets his house be used
Houses such that of two races either both are present or both absent	The two-person house in which members of these two races live
Houses such that of three races two or none are present	The three-person house in which members of these three races live
Houses in which an even number of people live	The four-person house in which one of each race lives

Choose any three septets from the above that have three houses in common. The three houses in which the parties will be held: do they form a triad? Is this so sometimes? Is it so always? Are the houses common to the three septets the same houses as the ones in which the three parties are held? Is it so sometimes? Is it so always? Is it never the case?

Another Game with the Fifteen Houses

On rainy days during the monsoon season the islanders liked to solve puzzles. One day they made models of their houses and wondered how they would have placed their houses if their island had been rectangular instead of pentagonal. They decided that they would probably have made three rows of five houses (or five rows of three houses). They also thought that the differences between neighbouring houses (as far as their inhabitants were concerned) had to be decided beforehand. Here is their plan:

1 1 1 1
1 3 2 3 1
1 2 2 1
2 3 4 3 2
2 1 1 2
1 3 2 3 1

The numbers between the houses show in how many ways the populations of two neighbouring houses must differ.

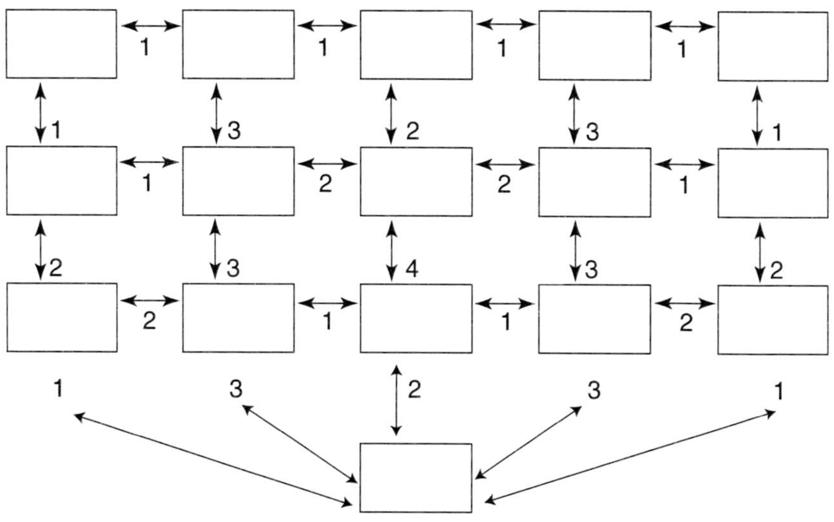

Figure 10

You can now play a triad game. Take two cubes. On one cube draw the houses that are in one of the columns, drawing the same house on opposite faces. Then choose another column and draw its houses on the faces of the second cube. You will have two dice. Throw both dice at once. Find the third house that will complete the triad which contains the two houses you have thrown with the dice. This you will find among the nine remaining houses in your village. The first player who correctly points out the third house of the triad that completes the two houses thrown by the dice gains a point.

You could also play the game backwards. Choose any two columns, which would be the houses on the two dice, if you played the game with the dice. Pick any house which is not in either column. Then find a house in one of your chosen columns and another one in the other of your chosen columns, such that the three houses together form a triad.

Another line of enquiry might be this. Look at one of your columns. See how the patterns change as you move from the first row to the second, then from the second to the third, then from the third back to the first. Write these changes down as a cyclic change. This means that if you started the same cycle somewhere else, not in the first row, and you had the same succession of changes, you could think of the cycle as the same cycle.

According to the above, do all the columns have the same cycle?

You will find one solution to the problem in Figure 11. There are lots of others. The English are shown with upright lines, the Africans with level lines, the Chinese are shown with small squares and the Indians are shaded grey.

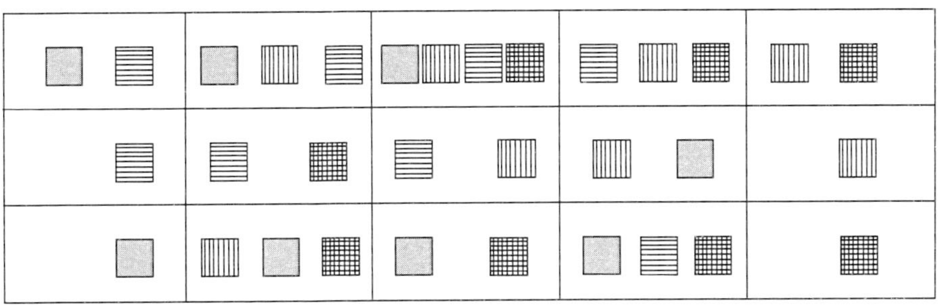

Figure 11

Let us denote the Eastern races (Indians and Chinese) by means of the number 1, the Western races (African and English) by means of the number 2, and if there is an Indian with an African or a Chinese with an Englishman, we shall write 3. The Indians and the Africans are the dark races; the English and the Chinese are the light races. To symbolise a house we shall write the dark inhabitants first and the light ones second (see Table 6).

Table 6

30	32	33	23	03
20	21	22	12	02
10	13	11	31	01

Note the symmetry about the vertical line through the centre. Note also that in all columns the three numbers 3, 2 and 1 behave cyclically, and in the same sense.

This means that we have the cycles for the dark inhabitants as well as for the light inhabitants (see Figure 12).

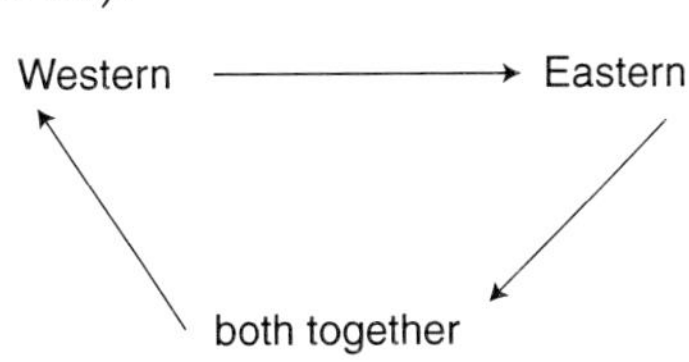

Figure 12

In the first column only dark persons come in and in the last column only light persons come in. If you found another solution, would you still find such cycles? Might each cycle involve more than two kinds of people?

You might have noticed that if houses A, B and C form a triad, then the sum of any two of the houses is equal to the third house. By sum we should understand the house in which people of those races live which are represented in either one or the other of the houses but not in both. We can then work out our addition table, as shown in Table 7.

Table 7

+	0	1	2	3
0	0	1	2	3
1	1	0	3	2
2	2	3	0	1
3	3	2	1	0

It is easy to see how the table works. If you add a house to itself, there will be nobody living in only one of them, so the sum is zero. If you add a house to the zero house (the clubhouse, where nobody lives), you will get, as the sum, the same house. Then out of 1, 2 and 3, any two will always yield the third one as the sum. This is easy to see since Eastern is 1, Western is 2 and both together are 3 and:

West + East = West–East
East + East–West = West (as East is in both houses)
West + East–West = East (as West is in both houses)

Now, what about multiplications? These could be the application of the cycles. If one of the factors is 0, the product should be zero. If one of the factors is 1, the product should be the house which is the other factor. If one of the factors is 2, then if we keep applying this factor on the other one, we should get the cycle:

$1 \rightarrow 2 \rightarrow 3 \rightarrow 1$

as our products.

If one of the factors is 3, then if we keep applying this factor on the other one, we should get the cycle:

$1 \rightarrow 3 \rightarrow 2 \rightarrow 1$

as our products.

So we have the following multiplication table (see Table 8).

Table 8

X	0	1	2	3
0	0	0	0	0
1	0	1	2	3
2	0	2	3	1
3	0	3	1	2

At one meeting of the islanders they decided to have an emblem for each house. They decided that each house would be represented by a certain position of the body. For this they thought the arms would be the most useful. They thought of the following possibilities:

(i) horizontally outstretched arm (for 0)
(ii) horizontal arm bent at the elbow (for 1)
(iii) arm placed vertically upwards (for 2)
(iv) arm vertical but bent over head at elbow (for 3)

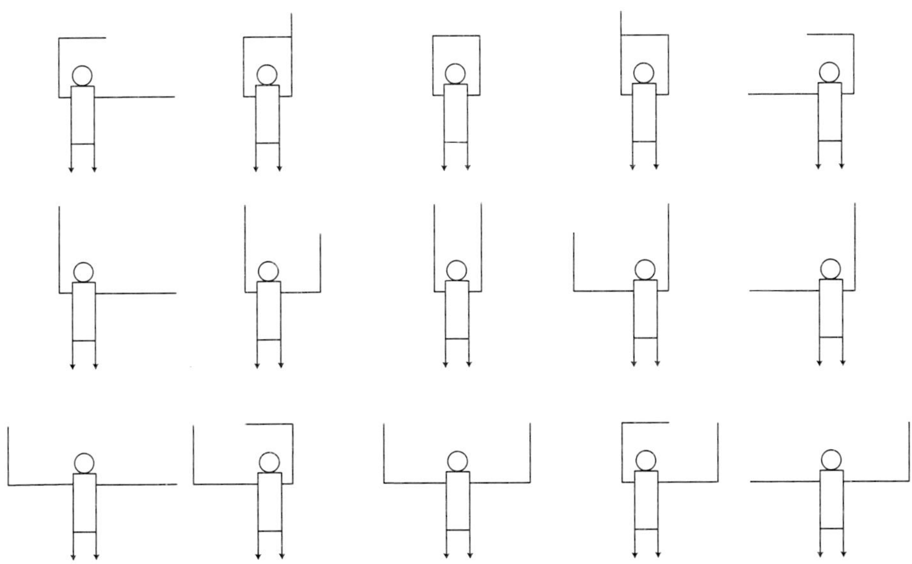

Figure 13

You can see that 'multiplying by 2' means going through the cycle:

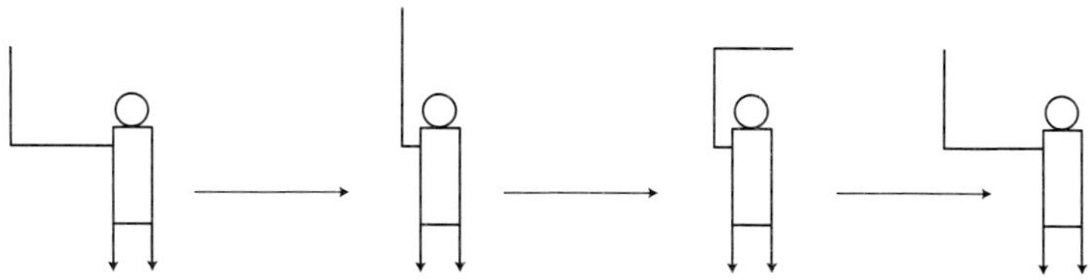

Figure 14

and multiplying by 3 means going through the cycle:

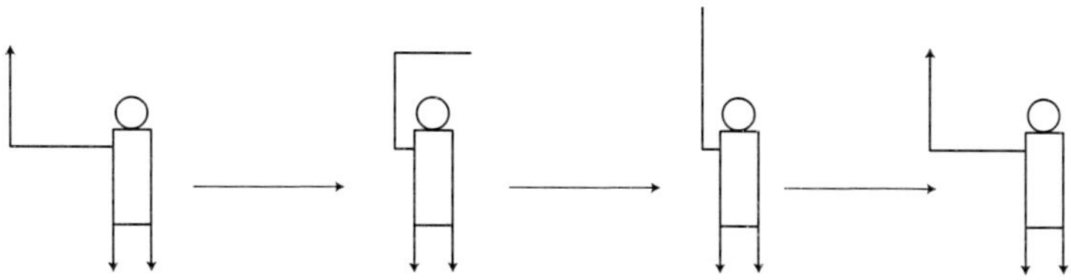

Figure 15

The positions given in the third row of the first and of the fifth columns are the basic positions. These are 10 and 01. All others can be obtained by multiplying and adding, starting from these two positions. Let us call these positions X and Y respectively. Suppose we wanted to obtain 23. All we have to do is to multiply X by 2 and Y by 3 and add the results.

We shall obtain 23, as shown in Figure 16.

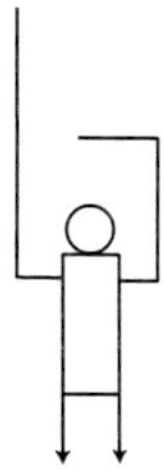

Figure 16

Can we do the same thing if we choose two different positions as basic positions? The answer is yes, as long as we do not choose both positions from the same column. In other words one of our positions should not be a multiple of the other!

This would make a fun game. One player chooses two basic positions and asks his or her opponent to find a way of calculating some other position chosen by him or her. The players then change roles. The player with a greater number of correct calculations is the winner. Here is an example:

The first player chooses the positions 11 and 02 as shown in Figure 17.

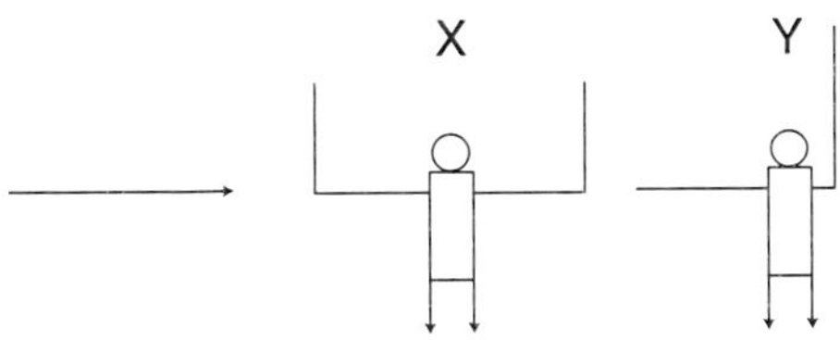

Figure 17

The opponent must calculate how to get 10, namely the figure here (see Figure 18):

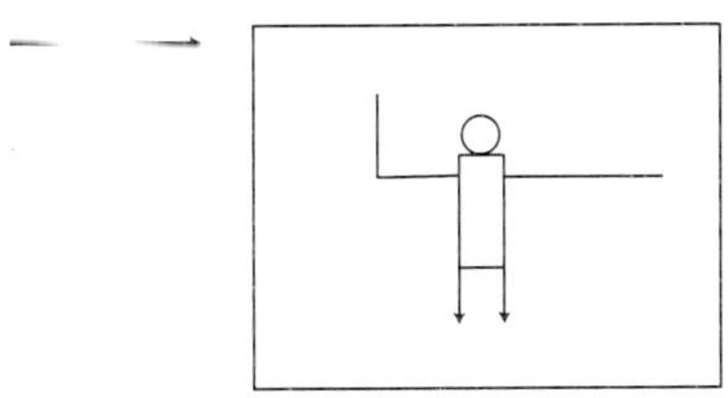

Figure 18

But this is easy; what we must do is

1X + 3Y

since

1X = 11 and 3Y = 01 and 11 + 01 = 10

(not forgetting that 1 + 1 = 0!)

Now choose your own basic positions and see if you can calculate all the other positions in terms of your chosen basic ones. Do not forget that when we multiply, we operate on the left and on the right arms. When we add, we add left arm to left arm and right arm to right arm.

Here is another way of arranging the houses. In this arrangement you have to make sure that in any seven houses marked by the same letter, members of a septet should be placed. Here is the map:

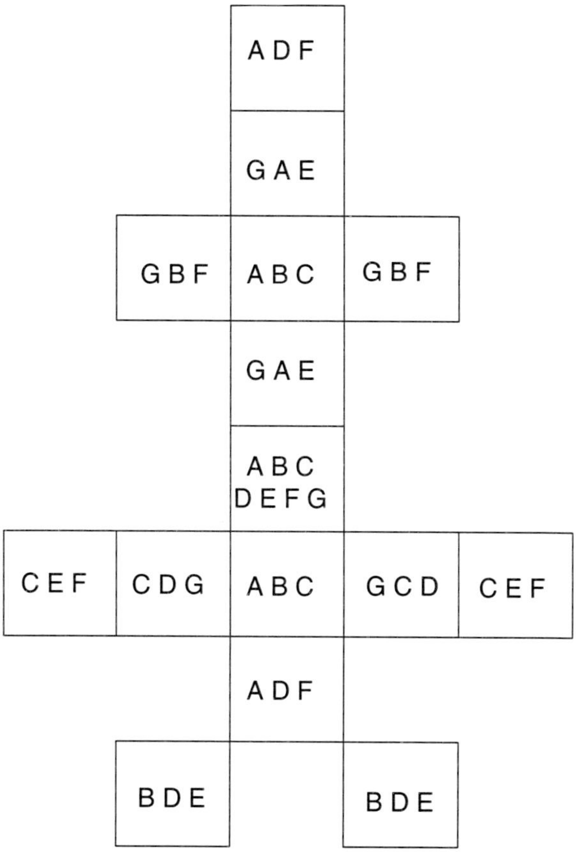

You can have a solution with septets arranged like this (the = sign means either both or neither):

A Even number
B Indians = Chinese
C Africans = English
D Chinese = Africans
E Indians = African
F Indians = English
G Chinese = English

The septets shown are all placed symmetrically about the line down the middle. Try to find them. In the diagrams given further on you can identify them!

On another rainy day they wondered how they would have built their houses on a circular island. They all thought one big circle would be the obvious solution. When they tried to arrange the triads round the circle so as to have them distributed evenly, the best they were able to do was to arrange them in such a way that out of any five consecutive houses, the first, the second and the fifth had to make a triad. Somebody then noticed that if they did that, then in a run of eleven houses round the island the first, the second, the third, the fifth, the sixth, the ninth and the eleventh houses formed a septet.

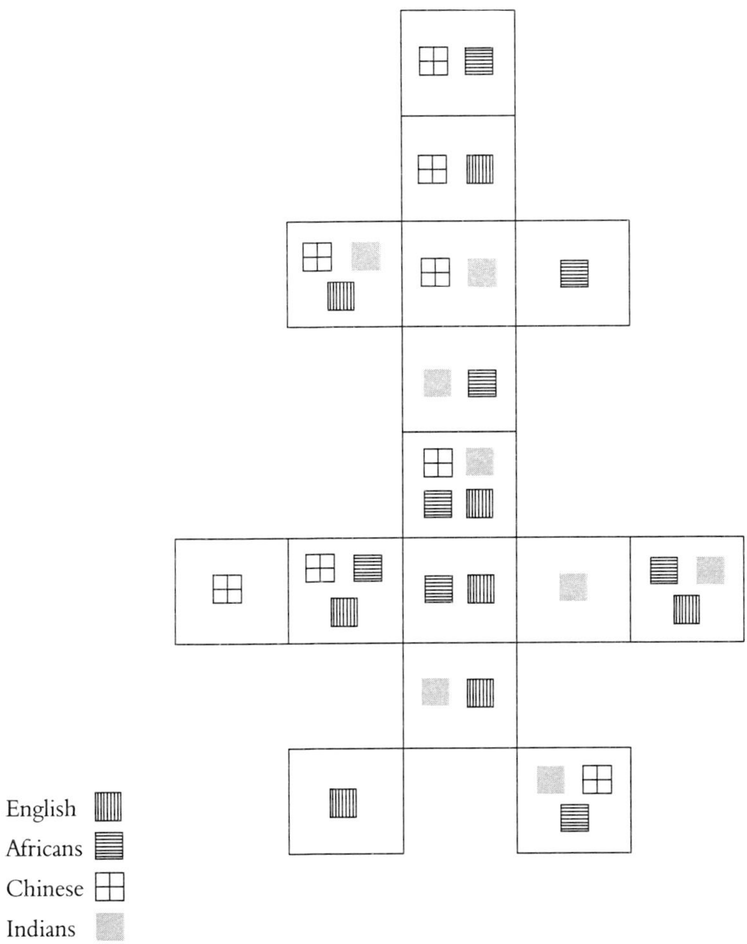

Figure 19

Here is how one group working on the problem arranged their model houses (see Figure 20).

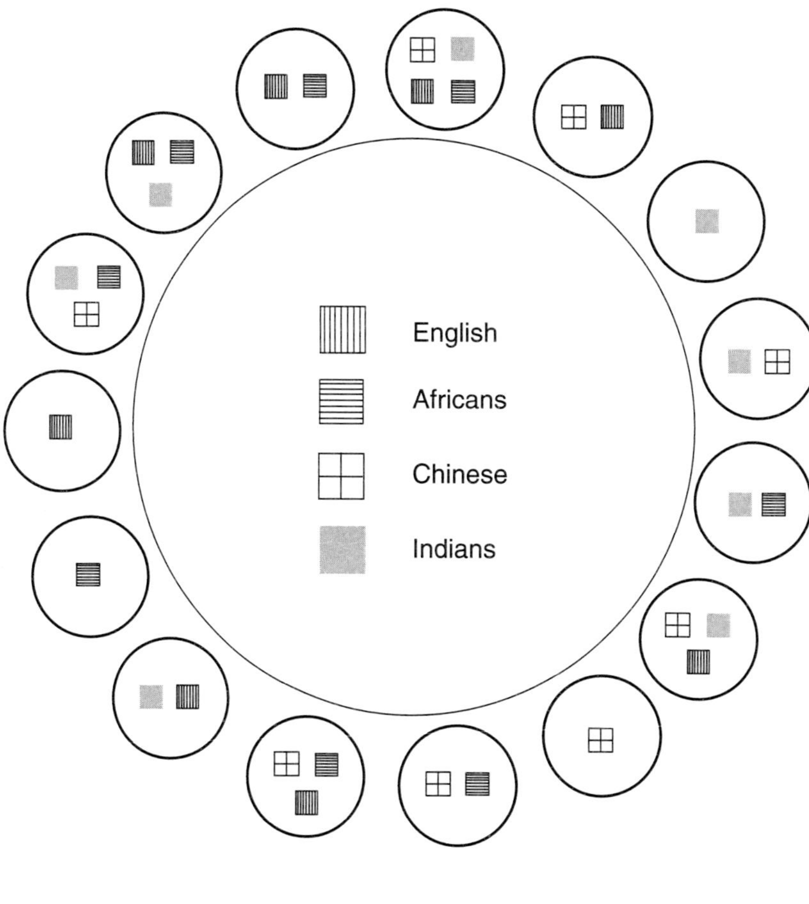

Figure 20

The rule for knowing who lives in the next house in the clockwise sense of the circle is given inside the circle in Figure 21 on the next page. If in a house you have one or three of the races shown before the arrow, then there will be a person of the race shown after the arrow in the next house.

You could now try to find the rule for the next house in the counterclockwise sense. Or for the next house but one!

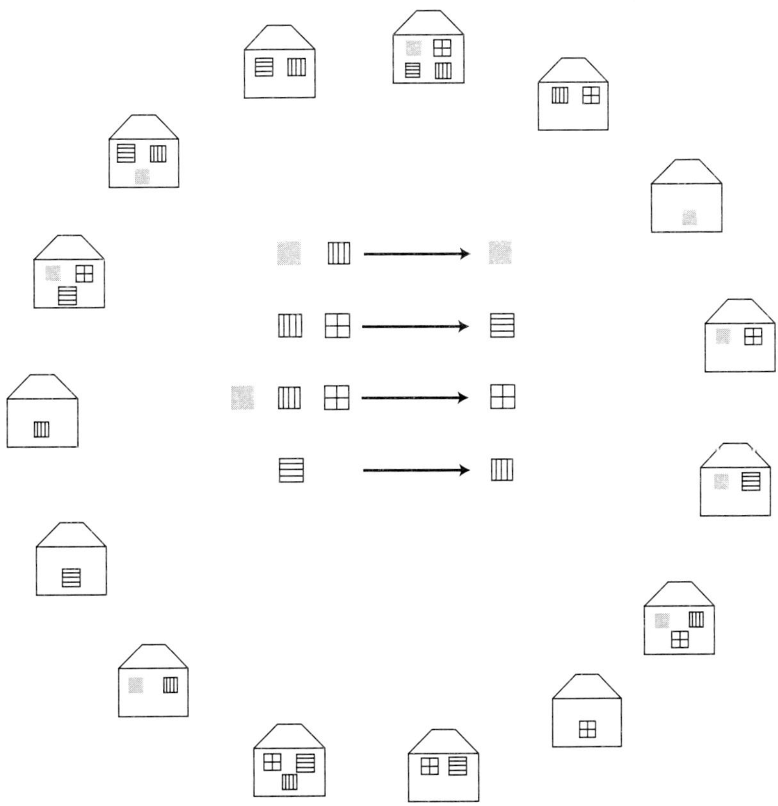

Figure 21

Here are some diagrams to show some possible septets in the first problem given.

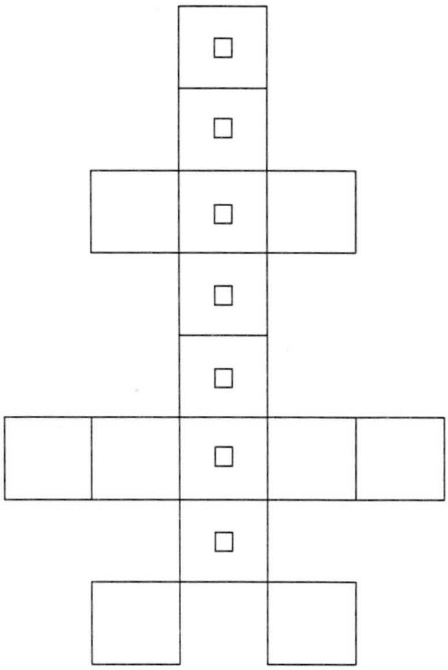

A(even number)

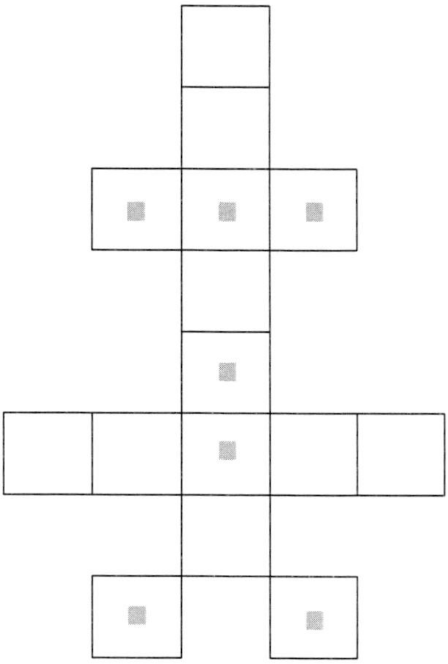

B(Indians = Chinese)

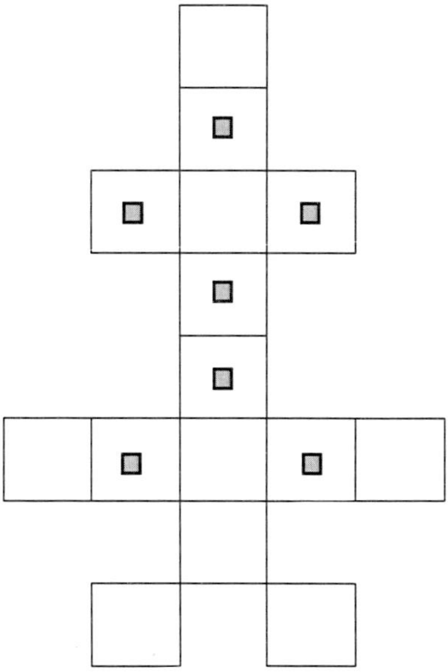

G(Chinese = English)

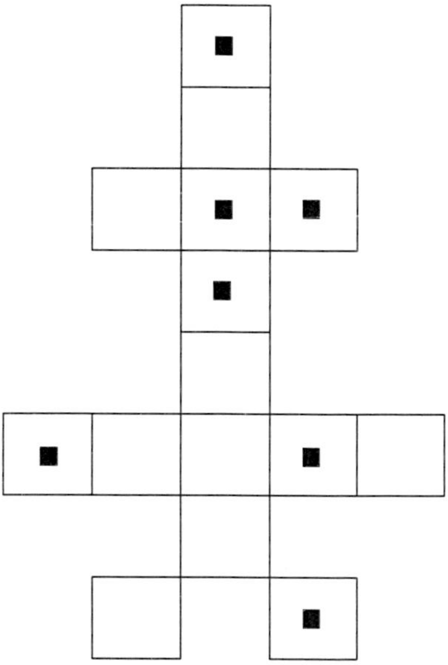

(No English)

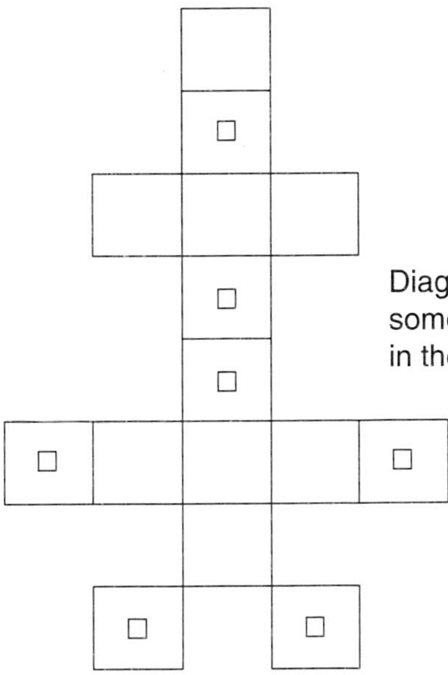

E(Indians = Africans)

Diagrams to show some possible septets in the first problems given.

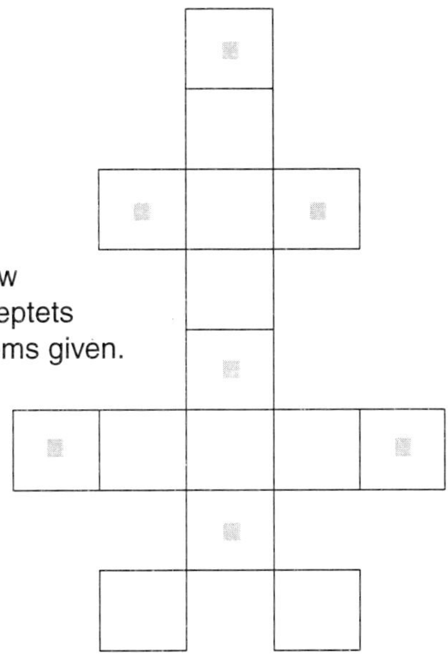

F(Indians = English)

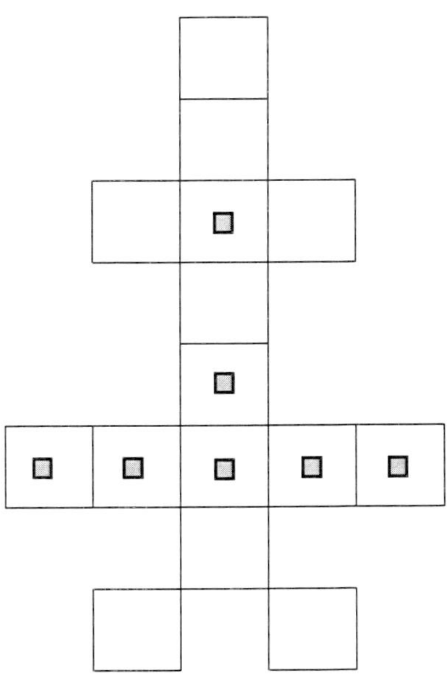

C(Africans = English)

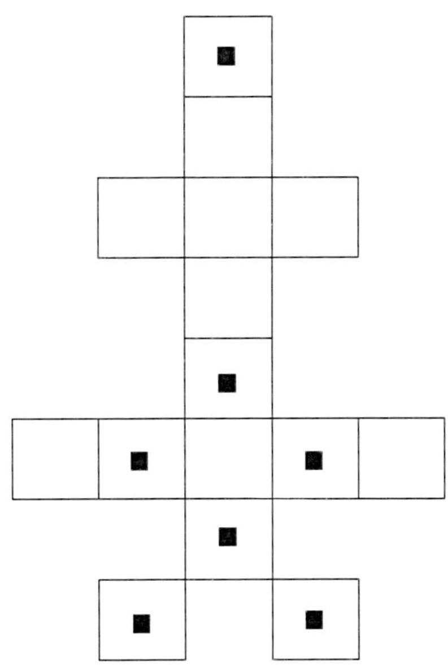

D(Africans = Chinese)

Figure 22 shows how they thought they would live on a triangle.

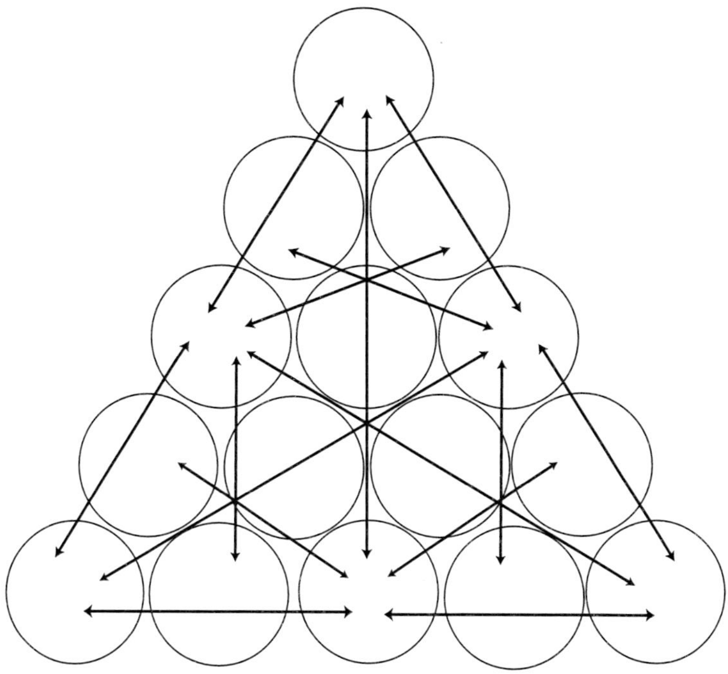

Figure 22

The arrows show how they wanted to arrange their triads. The circles in the middle, housing the three central houses, also mean that these three houses also should form a triad. Make your towers with coloured cubes and try to arrange them so that any set of three houses joined by an arrow should form a triad.

Figure 23 shows one solution for the triangle and an arrangement for the blocks, a puzzle for you to work out!

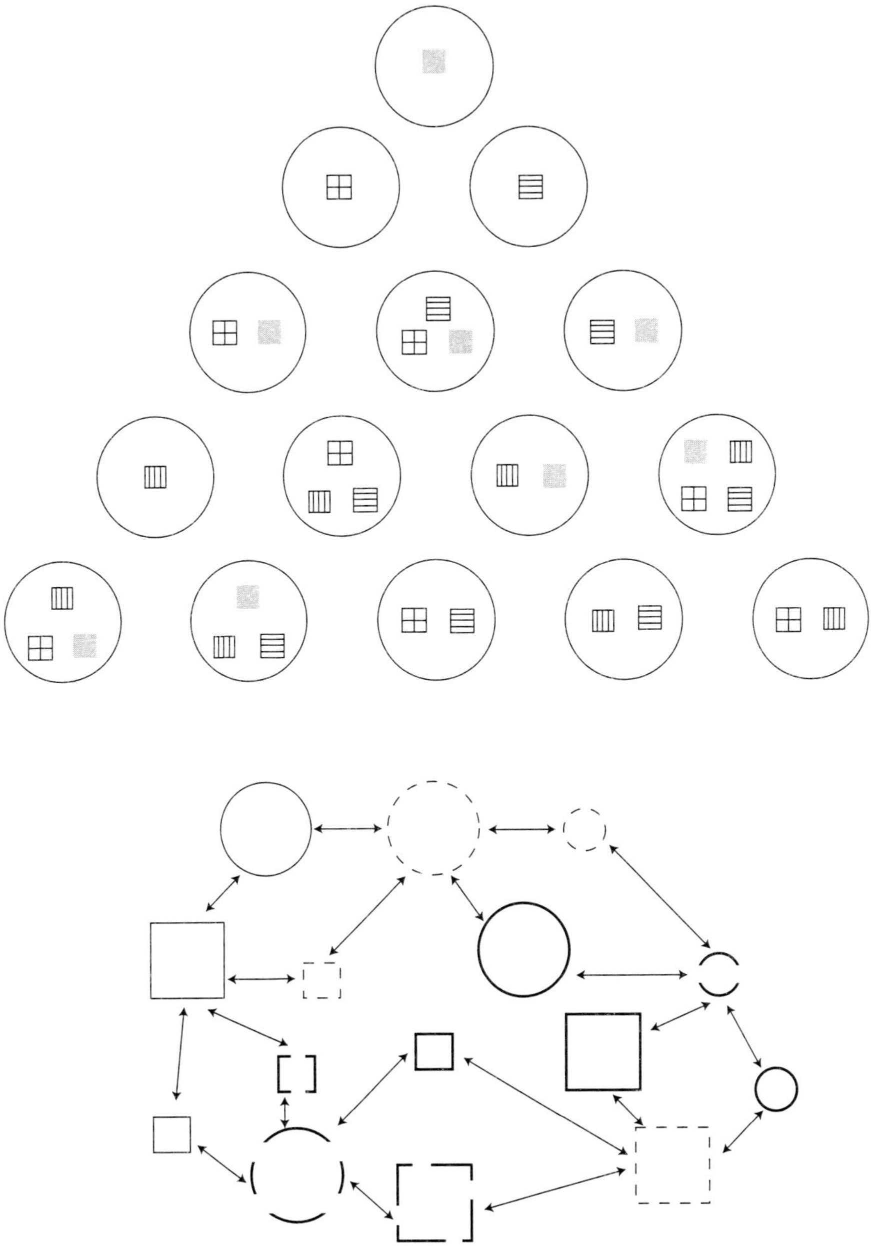

Figure 23

What mathematics have you learned? The following is strictly for the mathematically curious!

The fifteen houses, together with the clubhouse, represent in concrete form what in mathematics is called a vector space. The houses represent the vectors. Making the triads is an introduction to what is meant by a subspace. But to understand what a subspace is, we have to be aware of how to add two vectors. Adding and finding the third house to make up a triad come to the same thing in this particular representation. Adding a house to another, we get the house with those races in it which are in either one or the other of the houses to be added but not in both. The clubhouse represents the zero vector.

To symbolise any particular house, we could decide to write 1 for a race present in a house and 0 for a race absent in it. These numbers 0 and 1 are called scalars. We still have to decide the order in which we list the possible inhabitants. We could take the order red, blue, yellow, green, where red would mean Indian, blue African, yellow Chinese and green English. Then we could write:

1100 for red and blue
0111 for blue, yellow and green
0000 for the clubhouse

and so on.

The addition rules for our scalars (namely for each colour or race) would be:

$$0 + 0 = 0, 0 + 1 = 1, 1 + 0 = 1, 1 + 1 = 0$$

the last one following from the fact that in adding you do not put in the race that is present in both houses.

So every vector would be written in four parts (called components), each part referring to a different colour. To add two vectors, we add the colours separately, red to red, blue to blue, yellow to yellow and green to green. Multiplying a vector by 1, we obtain the same vector. Multiplying by 0, we obtain the zero vector. We shall use the symbol ⋆ to denote multiplication.

The sum of any two vectors is always the third vector that makes up the triad determined by the two vectors to be added. If A, B and C are vectors forming a triad, then:

$$A + B + C = 0 \text{ and } A + B = C$$

Any vector V determines a set of two vectors, itself and the zero vector, since

$$1 \star V = V \text{ and } 0 \star V = 0000$$

Any two non-zero vectors U and V determine a triad since

$$1 \star U + 0 \star V = U, 0 \star U + 1 \star V = V, 1 \star U + 1 \star V = U + V$$

and of course

$$0 \star U + 0 \star V = 0000$$

since the zero vector, invisibly, belongs to every triad.

Any three vectors not forming a triad, determine a septet. A septet has eight members, if you include the zero vector. Any four vectors not all in the same septet are enough to obtain all the vectors by multiplying by 0 or 1 and adding.

A set of two vectors consisting of any vector and the zero vector is called a one-dimensional subspace. That is because we need just one vector to determine all the vectors in it. A triad is called a two-dimensional subspace, as we need two vectors to determine all the vectors in it. A septet is called a three-dimensional subspace. This is so since we need three vectors to determine all the vectors in it. Our whole vector space is called a four-dimensional space, since we need four vectors to determine all the vectors in it.

The number of dimensions of a vector space is the smallest number of vectors needed for determining all the vectors in that vector space by multiplying by scalars and adding.

A group is a set of elements in which a binary operation is defined, such that the carrying out of the operation on any two elements results always in an element of the set. There must also be a neutral element. Every element must have a unique inverse. Finally the operation should be associative. If the operation is commutative, the group is called abelian.

It is easy to see that every subspace, as well as the vector space itself, is an abelian group according to the above definition. Every vector is its own inverse.

If we define a second operation, which is distributive with respect to the first, if it has its own neutral element, if it is associative and every element except the neutral of the first operation has a unique inverse in this operation, then we have a field.

A shorter way of saying the above would be to say that there should be a second operation which forms a group, as long as we do not include the neutral element of the first operation. Also the use of the second operation which involves the neutral of the first operation should always result in the neutral of the first operation (the zero vector).

If the above is too complicated, just think of a field as a set of elements in which all the usual rules of arithmetic are true. The set of scalars {0, 1} forms a field, with the previously noted addition rules and the multiplication rules:

$$0 \star 0 = 0, 0 \star 1 = 0, 1 \star 0 = 0, 1 \star 1 = 1$$

In our problems, however, we were using another field. This other field has four elements, which we can represent by the symbols 00, 01, 10, 11, the addition rules being to add the left digit to left digit and right digit to right digit by the rules of the two-element field we have already seen.

The neutral of the multiplication can be the element 01. The elements 10 and 11 multiply another element by means of the cyclic rules:

⋆10: 01 → 10 → 11 → 01
⋆11: 01 → 11 → 10 → 01

If we take any true addition and take it round the cycle by repeatedly multiplying it, we always get another true addition. This is what is meant by saying that the multiplication is distributive with respect to addition. Here is what happens:

10 + 11 = 01
11 + 01 = 10
01 + 10 = 11

The second addition is obtained by multiplying the first one by 10; the third one is obtained from the second one by multiplying the second one by 10.

The left digit refers to a warm colour, the right digit to a cold colour. So multiplication by 01, or by the cold colour, is the neutral, which leaves things as they are. Multiplication by 10, or by the warm colour, takes you round the cycle:

warm → cold → warm–cold → warm

while multiplication by 11, or by both colours together, takes you round the cycle:

warm → warm–cold → cold → warm

In the symbols we used to describe the rectangle game, we used the symbols 0, 1, 2 and 3 instead of 00, 01, 10 and 11. The vectors will be symbolised differently since we have used double digits here, instead of the single digits in the description of the games. But we can easily pass from one notation to the other by remembering that:

Single digits	Double digits
0	00
1	01
2	10
3	11

and so a vector that was previously written as 12 will now be written as 01 10, or 32 would now be 11 10 and so on.

So our field now has four elements. We have four scalars. Each vector consists of two scalars, given in a certain order. Clearly the order is important, as the vectors 01 10 and 10 01, for example, are not the same vector.

We now need only two vectors out of which we can reach all other vectors by multiplying by these scalars and adding. So the same set of sixteen vectors has been organised into a two-dimensional vector space, but based on a four-element field. If we use the two-element field {0, 1}, then we need four vectors to determine all our vectors. So just listing a set of vectors does not really give us a vector space. We need to specify the field on which the vector space is to be based.

If any of this is not clear, you can still have fun with the games! Or you can read a boring book on linear algebra!

Chapter IV

THE SIXTEEN GAME WITH MOUNTAINS, FLOWERS, TREES, HOUSES AND NUMBERS

1. The Deck of Cards

The pictures shown in Figure 24 will be on one side of the cards in the pack:

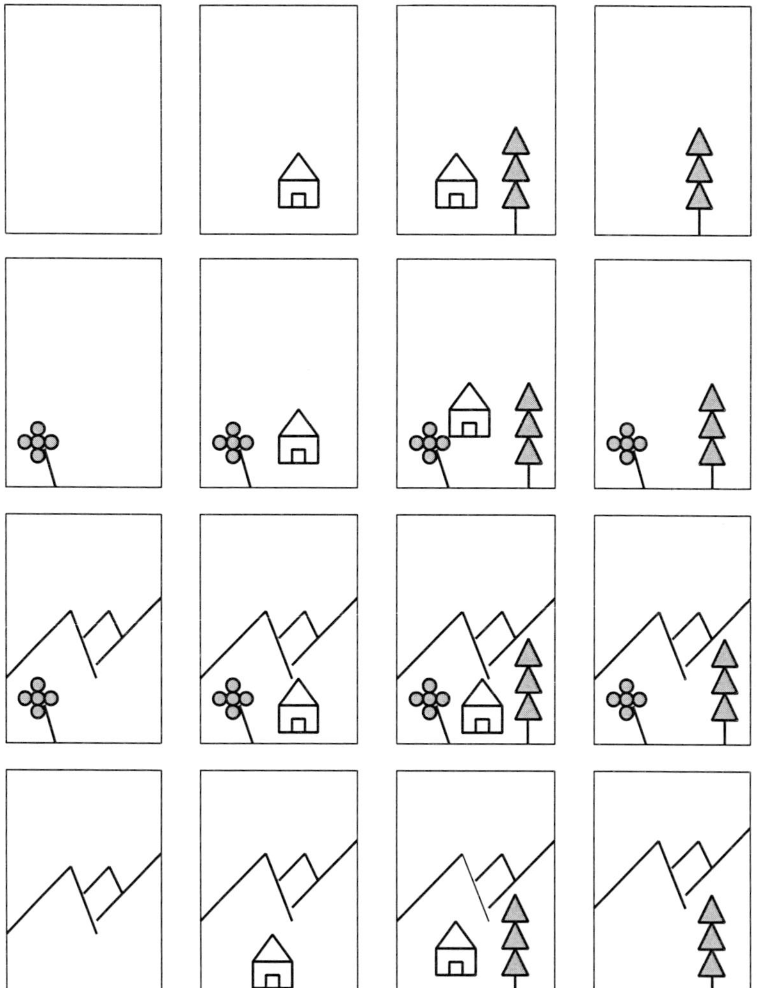

Figure 24

This is what will be on the other side of each card. The numbers are given in the same order as the pictures. Cut them out and stick them back to back. 'Full' means 210 circles! I just got tired of drawing them all!

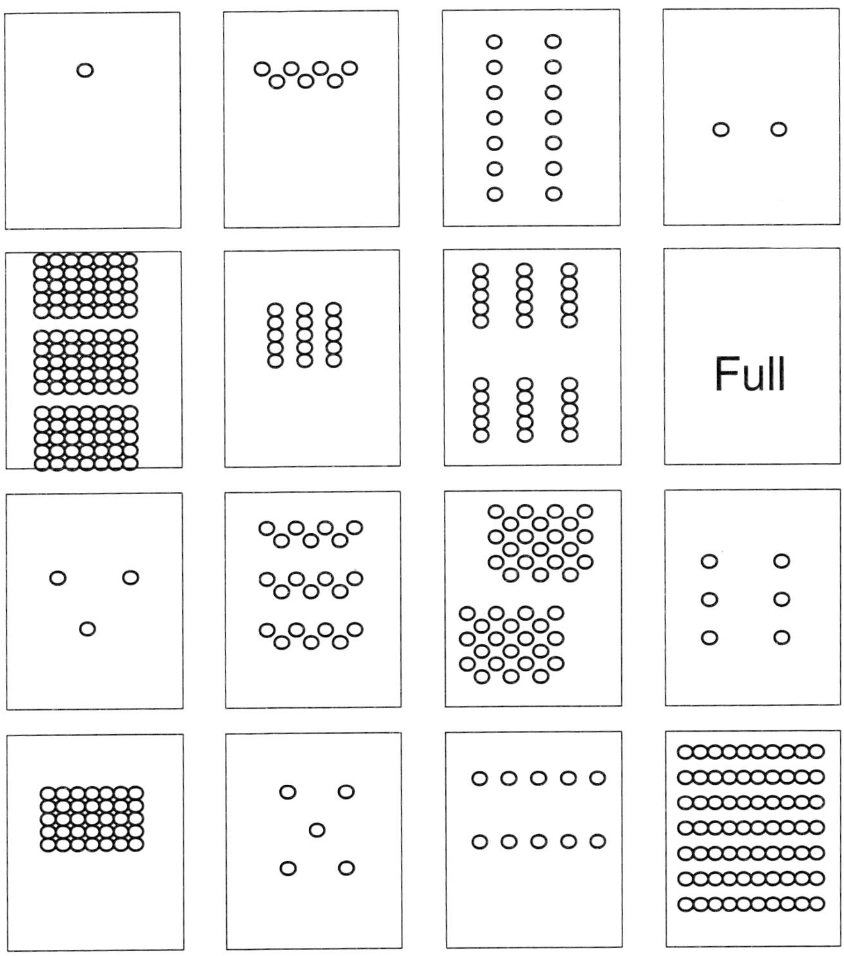

Figure 25

2. Guessing What Is on the Other Side

The deck is shuffled and each player is dealt five cards. An agreed amount of time is given for each player to study how one side of a card is related to the other side. Then each player places his or her cards down, with the pictures facing upwards. Each player tries to say what the numbers are that are on the reverse sides of the opponent's cards. The player who has the larger number of correct guesses is the winner.

The game can also be played by players placing their cards with the number sides facing upwards. Then players will have to guess what the pictures are on the reverse sides of the opponent's cards.

3. Solution to the Guessing Game

GUESSING THE NUMBER FROM THE PICTURE

Choose a card and look at its picture. The elements on the picture will lead to various numbers following the arrows below. Multiplying the numbers you get from the card will give the number on the reverse side of that card.

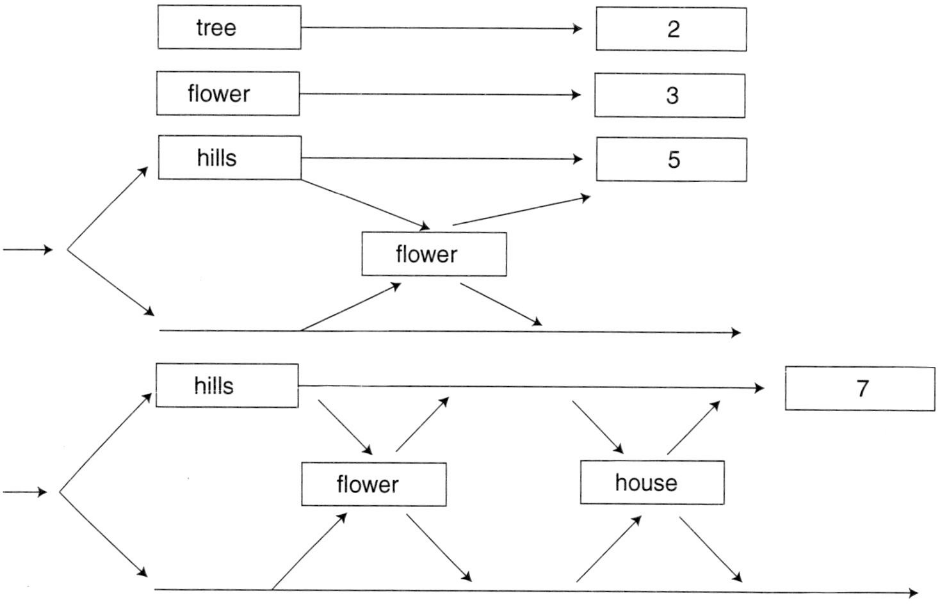

Figure 26

Follow the factors of your number through each track and you will find the elements of the picture on the reverse side of the card with that number.

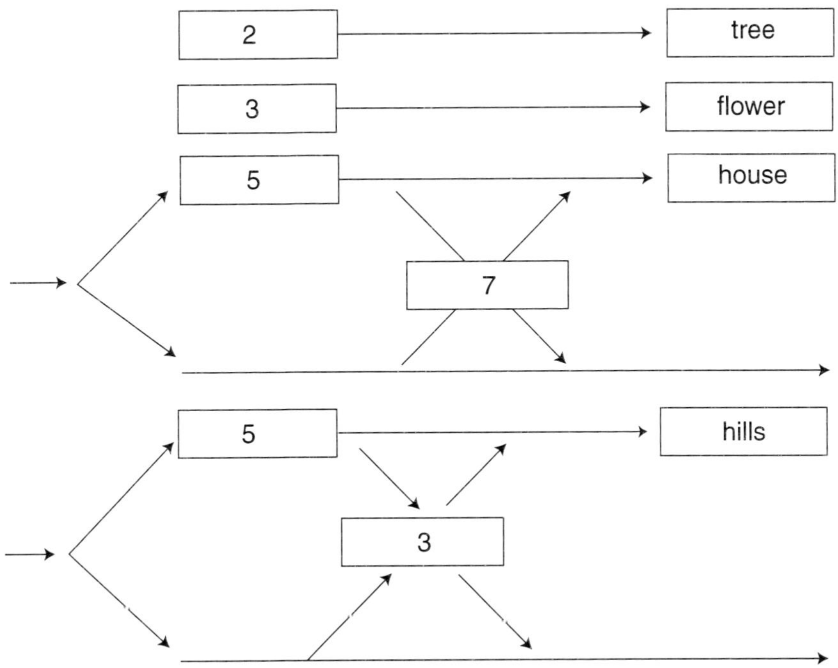

Figure 27

4. Finding Triads

You will know by now that all the numbers on the cards are factors of the number 210, which is the product of $2 \times 3 \times 5 \times 7$. The numbers 2, 3, 5 and 7 are the prime numbers out of which the game is constructed.

A triad is a set of three of our numbers on the cards such that each of the prime factors occurs either not at all or twice in the set of three numbers. For example, {6, 21, 14} is a triad since 2 occurs twice, 3 occurs twice, 7 occurs twice and 5 is missing.

A triad of pictures is a set of three pictures such that of each component there are either none or two in the set of three pictures.

Try to make as many triads as you can. There are 35 ways you can make a triad out of the fifteen numbers (not including the number 1) on the cards. If you make a picture triad, will the reverse sides give you a number triad?

If you make a number triad, will the reverse sides give you a picture triad?

THE FIVE TRIAD PROBLEM

Try to make five triads but such that no two triads have a picture or number in common. Arrange them as five columns of three pictures or numbers. Try to arrange your triads in such a way that in each row the first three numbers or pictures as well as the last three numbers or pictures will make triads.

If you have solved this problem for numbers by turning the cards over, will you have solved it for the pictures? Or the other way round? Solution below, but don't look!

THE FIFTEEN TRIAD PROBLEM

Place your pictures or numbers in the shape of a circle, but in such a way that in every run of five pictures or numbers, the first, the second and the fifth always form a triad. When you have done this, try to find the other twenty triads along your circle. You will find five of them symmetrically placed. How are the other fifteen triads placed?

A solution to the five triad problem is as follows:

2	3	6	5	30
7	105	15	42	70
14	35	10	210	21

5. Find the Rule for the Next Item Round the Circle

Here is one solution to the fifteen triad problem. There are many others!

2, 3, 5, 7, 6, 15, 35, 42, 10, 21, 30, 105, 210, 70, 14 and back to 2 again.

You can check that if you start anywhere and take every fifth number, you will again get a triad every time. These are the symmetrically placed triads. No doubt you will have located the others by now!

Now what would be the rule for telling what the next number is in the series? Do not read on if you want to find out by yourself!

There are three rules that stand out quite easily!

(i) After a factor 3 you always get a factor 5.
(ii) After a factor 5 you always get a factor 7.
(iii) After a factor 7 you always get a factor 2.

But how do you know whether the next number has a factor 3 or not? That is a bit less obvious!

It may not have taken you too long to find out that if a number has either the factor 2 or the factor 7 but not both, then the next number will have a factor 3. If a number has both these factors or neither of them, then the next number will not have the factor 3.

Of course, I have made my sequence as easy as possible; your sequence that you will have made for housing the fifteen triads as first, second and fifth out of any run of five might not admit of such easy rules. Go back to your solution and try to find the rules. If you have arranged your triads as required, there will be a rule to be found!

If you find difficulty, turn your cards over to the picture side. There will also be rules for which picture card comes after any picture card. For all you know, the picture rules might be easier to find!

If you are not too muddled up yet, then go to the next section and see what you can do with moving your arms around to represent what you have been doing with pictures and numbers.

6. Sixteen Ways of Holding Your Arms

Here are fifteen ways in which you can use your arms for playing the sixteen game; I have omitted outstretched arms both ways (see Figure 28):

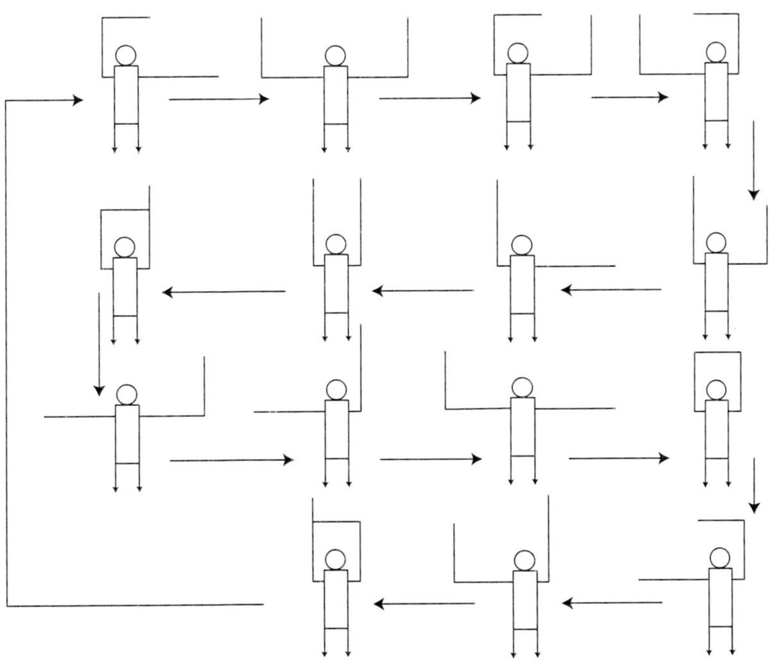

Figure 28

No doubt you will be wondering why I have placed these fifteen positions in this particular order. In fact, I had a rule in my head, according to which each position can only be followed by just one other position. The rule takes you round all the possible positions, except the one with two horizontally outstretched arms.

If you have not been able to gather my rule, here are some hints, which will help to describe the positions systematically:

(i) The left arm can be bent or straight. Bent will be called active, straight will be called passive. Code LB (meaning whether left arm is bent or not).

(ii) The left arm can be with the upper arm vertical or with the upper arm horizontal. Vertical will be called active, horizontal will be called passive. Code LV (meaning whether left arm vertical or not).

(iii) The right arm can be bent or straight. Bent will be called active, straight will be called passive. Code RB (meaning whether right arm bent or not).

(iv) The right arm can be with the upper arm vertical or with the upper arm horizontal. Vertical will be called active, horizontal will be called passive. Code RV (meaning whether right arm vertical or not).

Here are the 'rules':

(i) The arm on the left is bent in the next position if just one of LB and RV is active for the present position. If both LB and RV are active or if both are passive then the arm on the left in the next position is straight.

(ii) The upper arm on the left in the next position is vertical if just one of LB and LV are active for the present position. If both LB and LV are active or if they are both passive, then the upper arm on the left in the next position is horizontal.

(iii) The arm on the right is bent in the next position if LB is active in the present position. The arm on the right is straight in the next position if LB is passive in the present position.

(iv) The upper arm on the right is vertical in the next position if one or three out of LB, LV, RB are active in the present position. If two or none out of LB, LV and RB are active in the present position, then the upper arm on the right is horizontal in the next position.

If you feel like it you might check on the following assertion:

In any run of five positions, if you look at the first, the second and the fifth positions, there will be a total of two or zero active LBs, two or zero active LVs, two or zero active RBs and two or zero active RVs. Three such positions will form a triad.

Try to find all 35 triads!

Another, more pedestrian way of formulating the rules would be:

Present position	Next position
Left arm bent *or* right upper arm up	Left arm bent
Left arm bent *or* left upper arm up	Left upper arm up
Left arm bent	Right arm bent
One or three of: Left arm bent, left upper arm up, right arm bent	Right upper arm up

Whenever conditions in the left-hand column are not met, the opposite of what stands in the right-hand column will be the case. In other words, if an arm does not come out bent, it will be straight. If an upper arm does not come out up, it will be level.

7. Using the Numbers between 0 and 15

We can also use the numbers between 0 and 15 instead of the factors of the number 210. Each number can be made up using one or none of the numbers 1, 2, 4 and 8. This is how it can be done:

$$0, 1, 2, 2 + 1, 4, 4 + 1, 4 + 2, 4 + 2 + 1, 8, 8 + 1, 8 + 2, 8 + 2 + 1, 8 + 4, 8 + 4 + 1, 8 + 4 + 2, 8 + 4 + 2 + 1$$

So each number either does or does not have an 8, either does or does not have a 4, either does or does not have a 2, and finally it is either odd or even. We can call the numbers by the following names:

E number has an 8, F number has a 4, T number has a 2, and an odd number has a 1 in their expressions as sums of our allowed numbers.

You can then write the numbers on the reverse sides of our cards by following the rule given on the next page. Just take a card and notice what is on the picture. Then go through the 'tracks' as required by your card and note the number you reach. Add the numbers so obtained, and this will be the number on the reverse side of your chosen card. Do this for all the cards and you will have made another game of cards.

For example, if you take the card with the flower and hills and the house on it, you will get 0 for trees, 2 for the flower and 0 following the hills and the flower on the third track. You will get 8 following the hills, the flower and the house on the fourth track. So the number on the back of the card with a flower, hills and a house will be 2 + 8 = 10.

You can now work out what the tracks must be like for telling what is in a picture if you know its number. Or you could use these tracks backwards. For example, take 7 = 4 + 2 + 1. The 1 and the 2 tell us that there is a tree and that

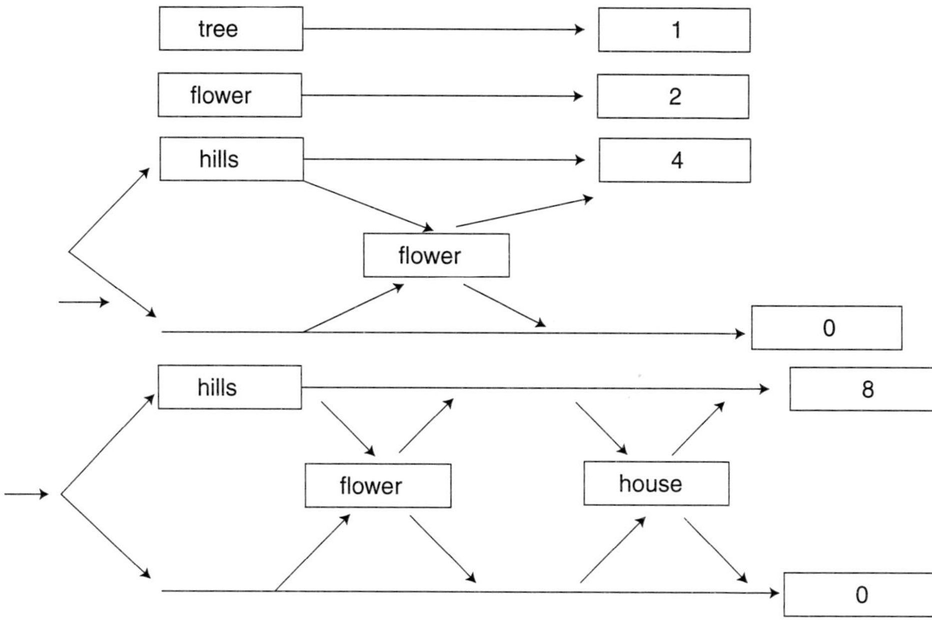

there is a flower. Since we have a 4 in 7, there cannot be any hills, because in the third track hills and flower would lead to 0. Finally the fourth track tells us that there cannot be a house, since with no hills and a flower the house would lead us to 8, but there is no 8 in our number 7. So the picture on the reverse side of 7 has just a tree and a flower on it.

Or we could take the number 9 = 8 + 1, which will have a tree but no flower by the first two tracks. The third track tells us that there cannot be any hills because hills and no flower would give us a 4 which we do not have in 9. Since there are no hills and there is no flower, the fourth track tells us that there must be a house, otherwise we would not have the 8 which we require. So the number 9 has a tree and a house.

Chapter V
TWO-DIMENSIONAL HOUSE–TREE GAME

1. The List of the Pictures for the Games

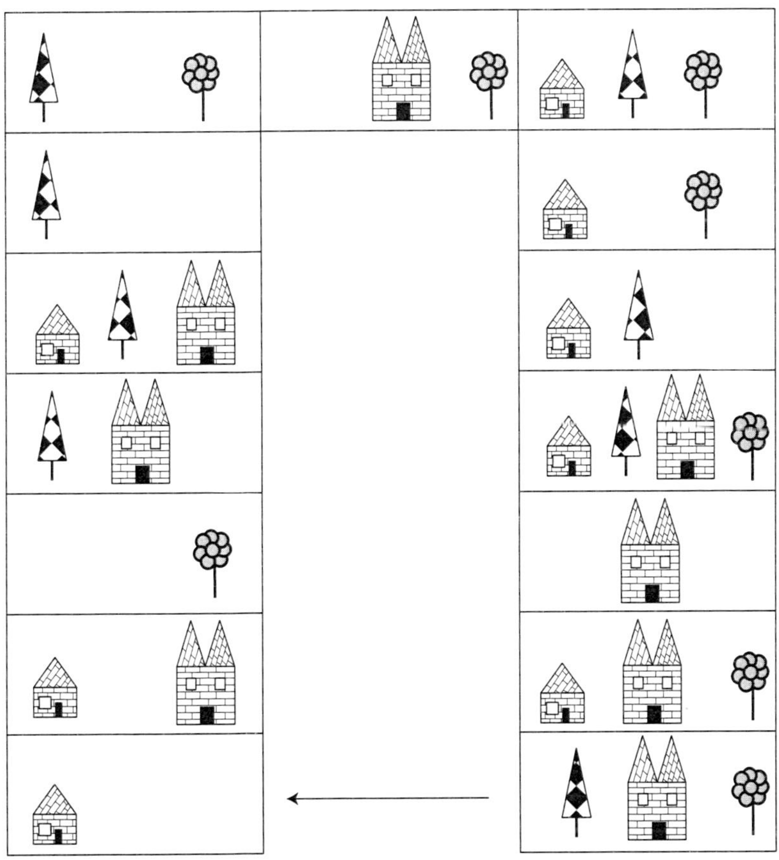

Figure 29

The pictures illustrated in Figure 29 are placed in a certain order. Can you discover the rule I used for putting the pictures one after the other?

The first dimension has the small house and the fir tree; the second dimension has the big house and the oak tree.

2. Arrangement in Columns

Going from left to right, we have these changes:

tree → house → tree and house → tree for each dimension

After the picture in the third column we come back to the picture in the first column. So the first column can be considered to be on the right of the third one (see Figure 30).

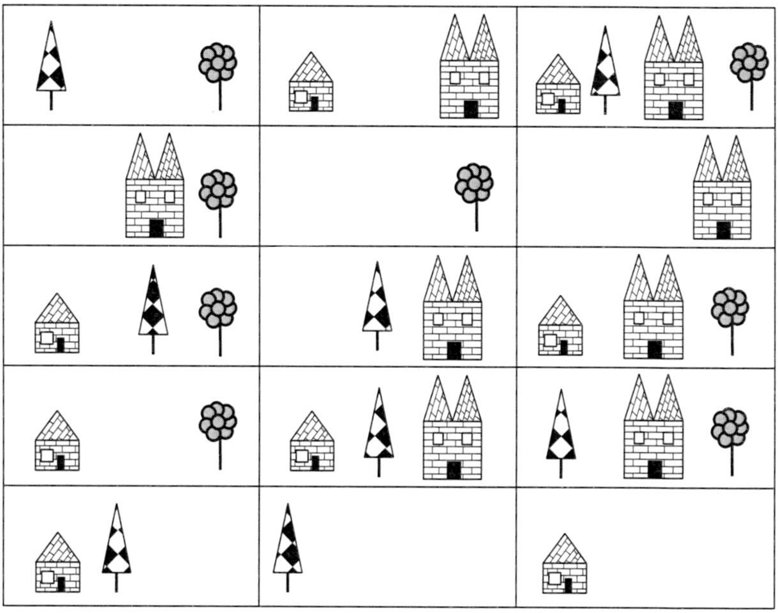

Figure 30

We can add trees and houses if we make the convention that two trees add up to zero, as well as two houses add up to zero. You will notice that in this case any two pictures in the same row will add up to the third picture in that row.

Choose any two of the five rows. Let me start with the last two rows.

Choose any picture from one of the other three rows. Find a picture in one of your chosen rows, which, when added to another picture in the other of your chosen rows, will add up exactly to the picture you have chosen from the other three rows. This will always be possible, whichever two rows you first choose and whichever picture you choose outside these two rows. Try to check that I am right!

3. Possible Sequence Rules

Now for the rule that gives you the next picture in the sequence. This means we are looking for a rule for getting the picture just below any given picture; in the case of the lowest picture in a row, below will mean the top picture of the next column on the left (see Figure 31).

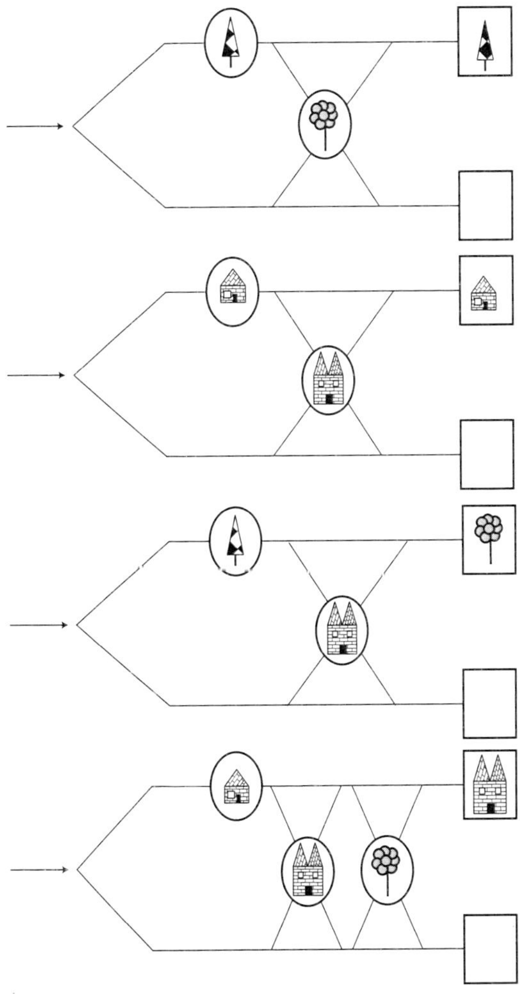

Figure 31

Here is one rule you can use:

To obtain what is in the first dimension of the picture below any given picture, just add everything in the picture you are looking at.

To obtain what is in the second dimension of the picture below any given picture, add the first dimension of the picture you are looking at to the second dimension of the picture to the right of the picture you are looking at.

Note that 'to the right of the third column' will mean 'the first column'.

Another way of telling what is in the picture below is through a flow chart.

The elements in the ovals refer to the picture you are looking at, the pictures in the rectangles refer to what you will find in the picture below. You will probably feel that the first method of looking at the dimensions is an easier option!

This second way of expressing the order of the sequence is like imagining that we have four dimensions instead of two. Each dimension now has only one object in it. But there is no reason why we should not have three objects in a dimension. Figure 32 is an example showing just one dimension:

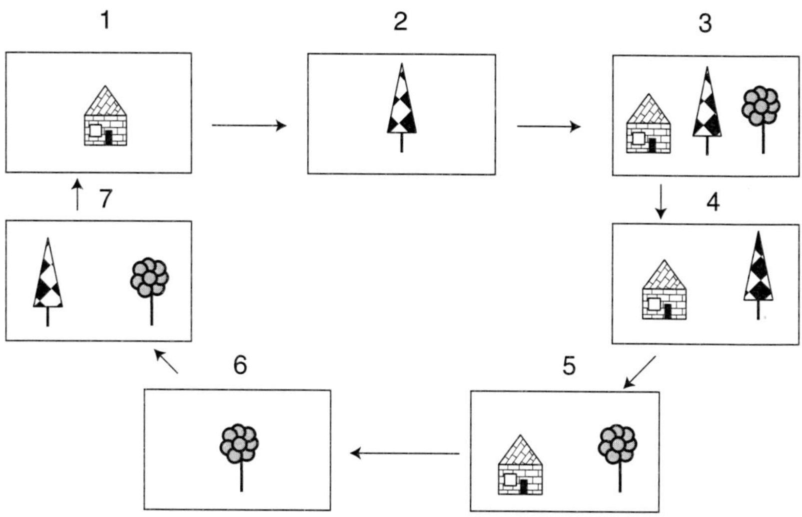

Figure 32

Putting any two of these pictures next to each other, we have two dimensions. We should also consider the picture with nothing on it, which could be the first picture or the second one. In this way we obtain 64 double pictures. Leaving out the double nothing, we are left with 63 double pictures.

4. Two Dimensions with Three Elements in each Dimension

Here is one way of listing all the possible double pictures. The numerals refer to the way the pictures are numbered in the previous section. The first numeral refers to the first picture, while the second numeral refers to the second picture of the double picture in question.

The adding of the elements is again defined as putting the elements together, and the sum of two of the same kind of element is zero. Adding is therefore done in the following way for the pictures (see Figure 33):

(i) Adding a picture to itself is zero.
(ii) Adding a picture to the next picture is two pictures on from the second picture.
(iii) Adding two pictures with one picture in between them, is the picture before the first one.
(iv) Adding two pictures with two pictures between them is the one after the first of the two pictures we are adding.

12	72	15	53	70	77	36	17	04
56	46	52	27	40	44	73	54	01
23	13	26	64	10	11	47	21	05
67	57	63	31	50	55	14	65	02
34	24	37	75	20	22	51	32	06
71	61	74	42	60	66	25	76	03
45	35	41	16	30	33	62	43	07

Figure 33

If you list the double pictures row by row, starting in the top left corner, then the rule for getting the next picture is this:

Let us call (A, B) any one of the double pictures, A being the one on the left and B being the one on the right. Then let us call (C, D) the next double picture in our sequence. C is obtained by adding A to the picture next to B in the cycle of seven pictures. D is obtained by adding A to the picture next but one to B in the cycle of seven pictures.

This is a 'full version' of the first column in our 9 × 7 rectangle of double pictures (see Figure 34).

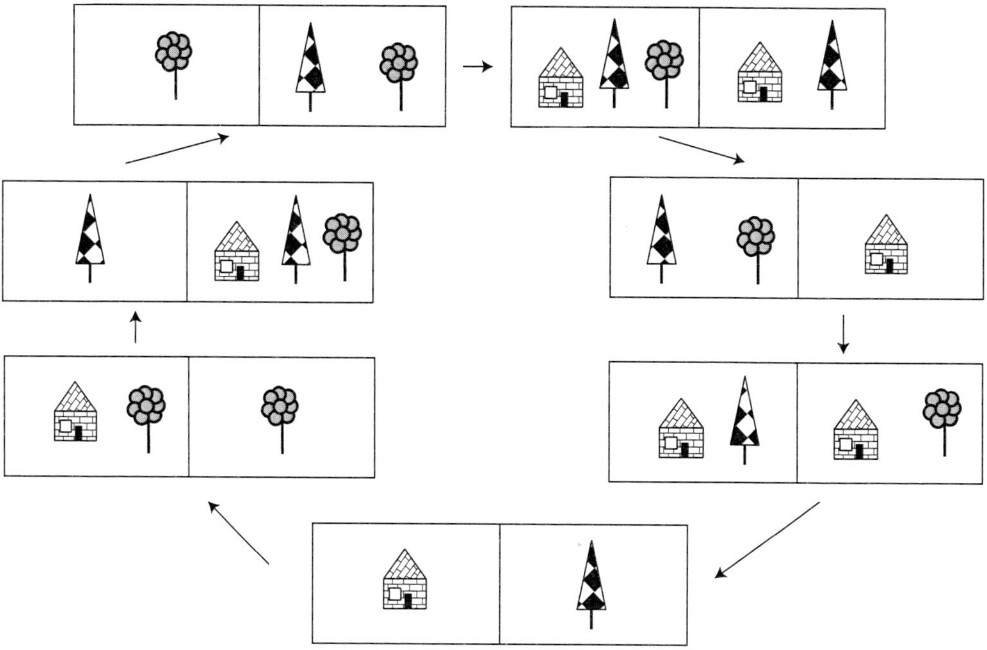

Figure 34

In the sequence shown in Figure 34 an arrow means four of the arrows of the sequence shown in Figure 35.

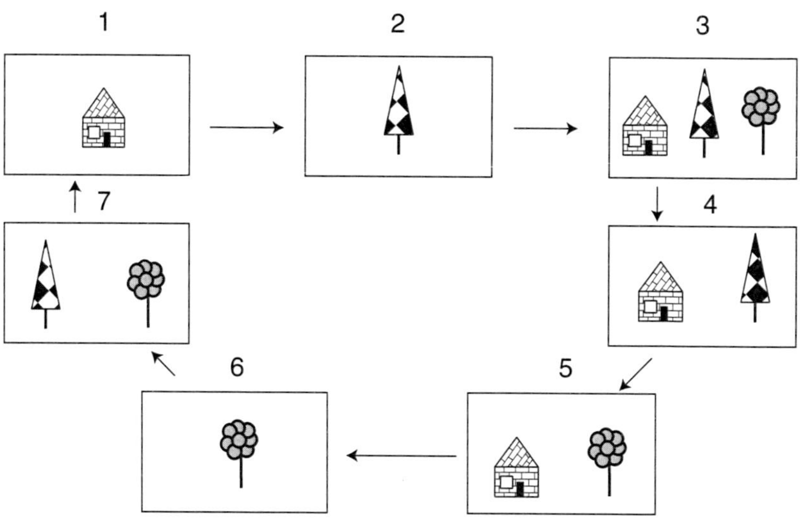

Figure 35

There are some possibly unexpected things we can discover in our 9 × 7 table. For example, the sum of any two double pictures right next to each other is easy to find. Just move back five double pictures from the first of the two you have added.

What about two double pictures with one double picture in between them? Take a diagonal step north-west from the first of the two you are adding. This means, of course, that if you want the sum of two adjacent pairs diagonally placed, you just have to move two spaces on from the double picture in the south-east.

If you take the next but one diagonally from north-west to south-east, then the sum will be four spaces on from the south-east double picture.

You can try adding double pictures that are a knight's move away from each other. Of course, you will get different results for knight's moves in different directions!

Another little problem that is amusing is this:

Choose any two columns in our rectangle of double pictures. Then choose a double picture not in either of your chosen columns. Find one double picture in one of your chosen columns and another double picture in the other one of your chosen columns, so that these two add up to the double picture you have chosen outside of your two chosen columns.

You will find that this is always possible, whichever two columns you choose, and whatever double picture you choose that is outside your two chosen columns. This is one reason why we call this game two-dimensional.

We could look at all six objects turning up in a double picture separately. Then we would have a six-dimensional game. Instead of two dimensions with three objects occurring in each dimension, we would then have six dimensions with one object occurring in each dimension.

It might be worth investigating whether it is possible to construct a six-dimensional set of flow charts to generate the sequence we have been considering.

Here are some rules you can check:

If in any double picture there is an odd number of the elements indicated in the first column below, then the next double picture will have the element shown in the second column below; in the case of an even number, the element shown in the second column will be missing in the next double picture.

(i) left house, right tree → left house next
(ii) left tree, right house tree and flower → left tree next
(iii) left flower, right tree and flower → left flower next
(iv) left house, right house tree and flower → right house next
(v) left tree, right house and tree → right tree next
(vi) left flower, right house →right flower next

Let us take two random consecutive double pictures as shown in Figure 36.

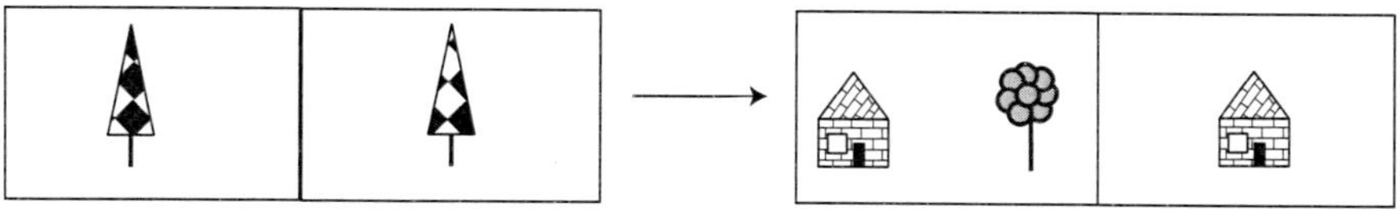

Figure 36

(i) Out of left house and right tree there is only one, namely a tree, which is odd, so we have a left house according to the first rule.
(ii) Out of left tree and right house, tree and flower there are two trees, which is even, so there is no left tree according to the second rule.
(iii) Out of left flower and right tree and flower we have just one tree, which is odd, so we do have a left flower, according to the third rule.
(iv) Out of left house and right house, tree and flower we again have just one tree, which is odd, so we do have a right house according to the fourth rule.
(v) Out of left tree and right house and tree we have both trees, which is even, so we do not have a right tree according to the fifth rule.
(vi) Out of left flower and right house we have nothing, which is even, so we do not have a right flower according to the sixth rule.

So in the next left picture we have a house and a flower, and in the next right picture we just have a house. Upon looking at the double picture which follows the two trees, we can verify that the rules work.

The six rules can also be expressed in the form of flow charts. Try to draw these charts from the given rules. If you find difficulty, look on the next page, where the charts are given.

Here are the flow charts which express the way we can calculate which double picture comes after any given picture. The right-hand end of the upper track of each chart tells us whether the element in question is or is not present in the following double picture. We must follow the chart from the input on the left, being guided by what we see in our chosen double picture whose successor we wish to find. If we arrive on the upper track on the right, the given element is present; if we arrive in the lower track on the right, the element in question is not present in the double picture shown in Figure 37.

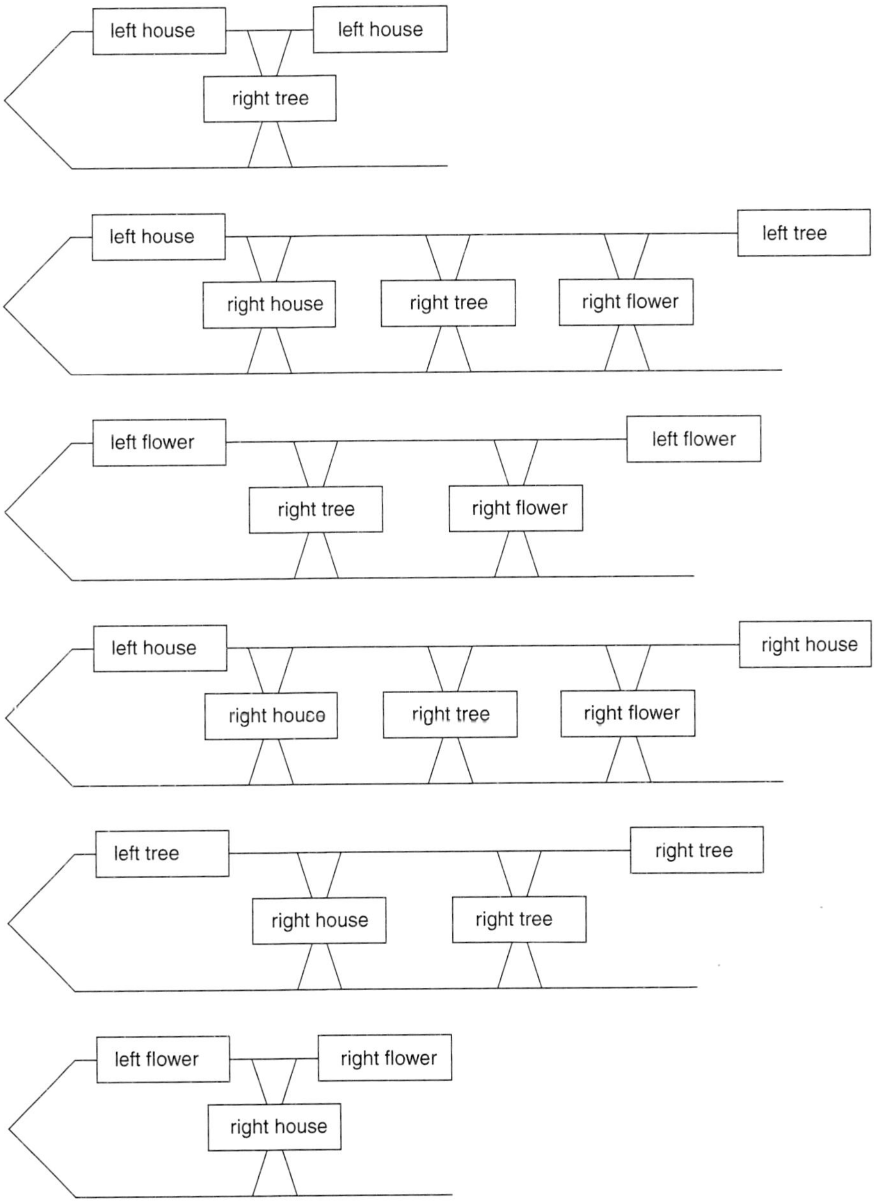
left house
left house
right tree
left house
left tree
right house
right tree
right flower
left flower
left flower
right tree
right flower
left house
right house
right house
right tree
right flower
left tree
right tree
right house
right tree
left flower
right flower
right house

Figure 37

Chapter VI
FOUR-DIMENSIONAL GAMES

1. Introductory Ideas

You will need some coloured or patterned counters in two shapes and four colours or patterns. Let us choose circles and squares in grey, level stripes, upright stripes and criss-cross pattern. These will be our toys.

A square and a circle of the same pattern will be called a zero pair. See, for example, Figure 38; these are all zero-pairs.

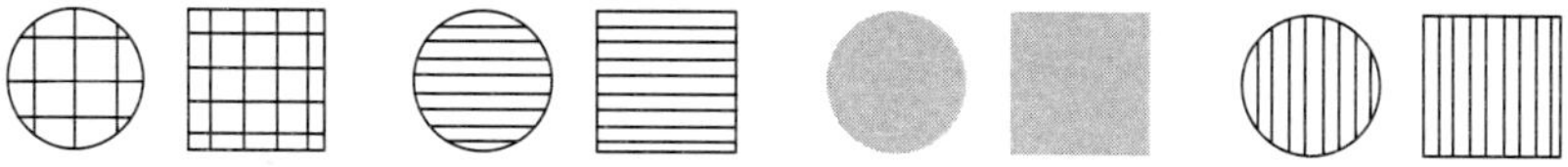

Figure 38

EQUIVALENCE OF TWO PILES OF COUNTERS

A pile is called equivalent to another pile if we can reach one from the other by either adding or removing some zero pairs (see, for example, Figure 39).

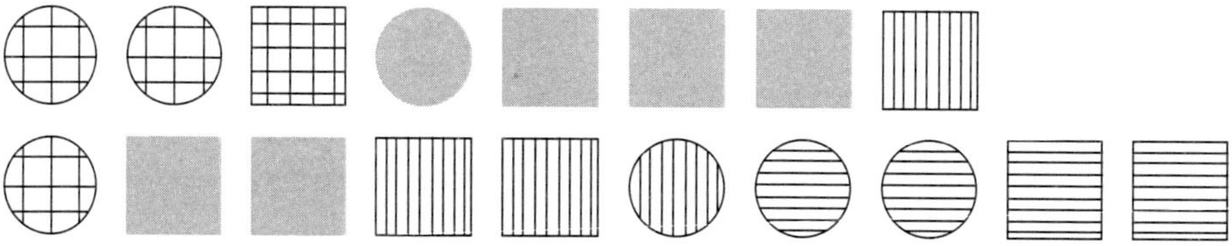

Figure 39

A pile of counters in which there are no zero pairs is called the standard form of the set of piles equivalent to it. In the case of Figure 39, the standard form would be as shown in Figure 40.

Figure 40

A pile of counters can always be replaced by another pile which is equivalent to it. Whenever possible, we shall use the standard form of an equivalent set of piles.

2. Addition and Subtraction of Piles

To add two piles we simply make one united pile out of the two piles. Usually this results in the appearance of some zero pairs; it is usual to take these away and so use the standard form.

A pile B can be subtracted from a pile A if the pile B forms part of the pile A. In this case we just remove the pile B, and the pile that is left is the pile A – B.

If pile B is not a sub-pile of pile A, then we can augment pile A by putting some zero pairs into it. We can go on putting in these pairs until pile B becomes a sub-pile of the pile A. Then we can take away this pile B and so still able to calculate the pile A – B.

Here is an example of an addition:

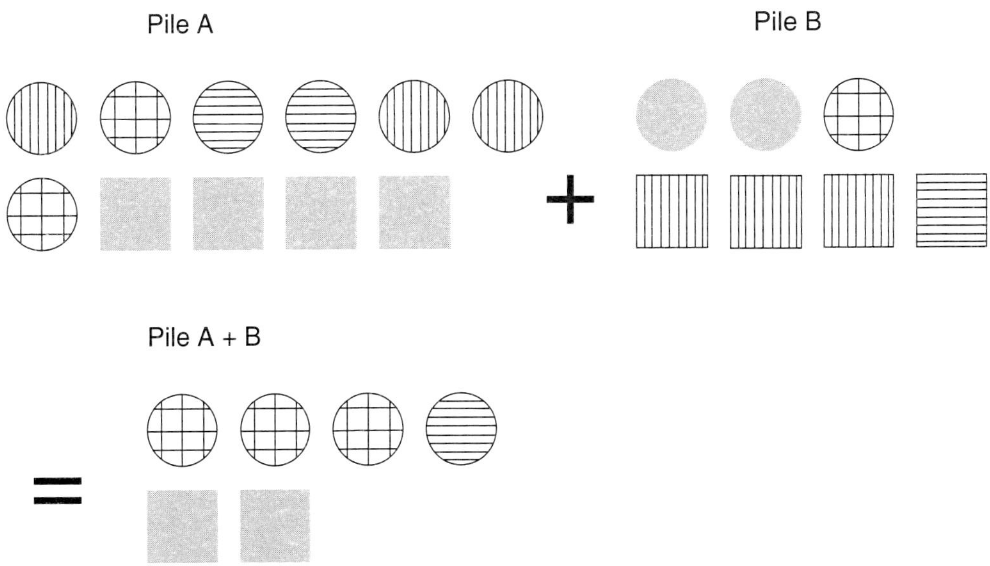

Here is an example of a subtraction which you can do as an exercise (augment pile A by putting in two grey squares and two grey circles):

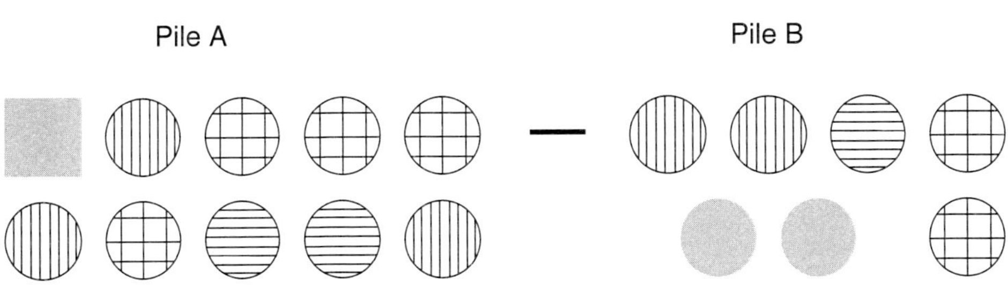

3. What About Multiplication?

You may have your own ideas about how to define a multiplication of a pile by another pile. It would certainly be reasonable to assume that one pile should be the multiplicand pile, and that the other pile should be the multiplier pile. Each counter in the multiplier pile would then act on each counter in the multiplicand pile. How this action should take place will depend on you. There are eight different kinds of counter. You will have to decide how each of these will act on each of the eight different counters.

If you have some ideas on how to do this, do not read on any further, but make up your own rules and see how they work out. If, however, you are open to suggestions, I can suggest two different ways of making up the rules.

SYSTEM 1

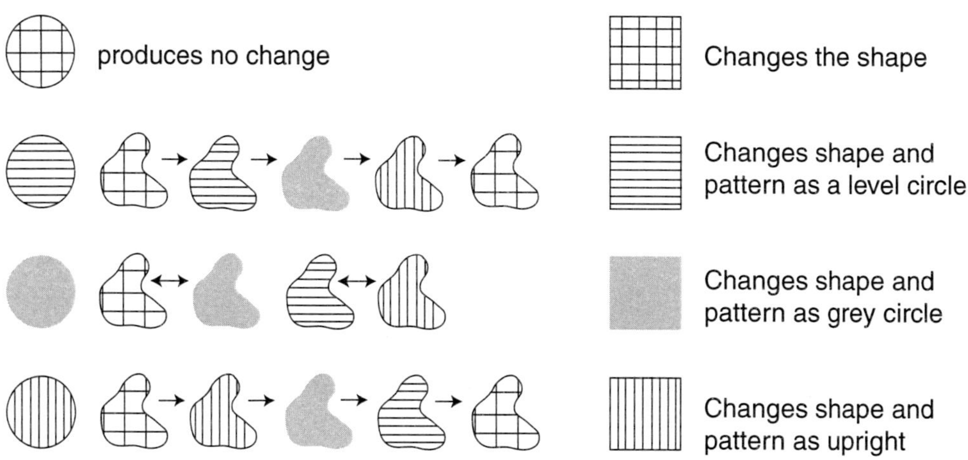

We can put a multiplicand pile in a box we shall call box A, our multiplier pile into another box that we can call box B, and then work out the product pile that we can place in a third box, called box C.

Here is an example of a multiplication:

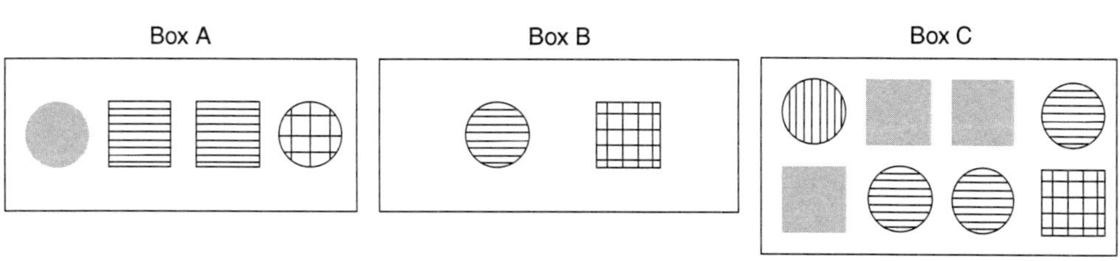

We obtain box C as shown since the level circle in box B will act on the counters in box A by turning criss-cross into level, level into grey and grey into upright and so produce:

The criss-cross square in box B changes the shape of what is in Box A and so it will produce:

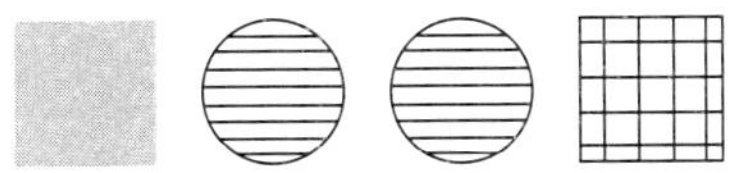

And, putting these two lots together, we obtain what has been placed in box C.

Here is another example, but now you find what you have to put in box B which will act on box A so as to produce box C. Do not look below if you want to try!

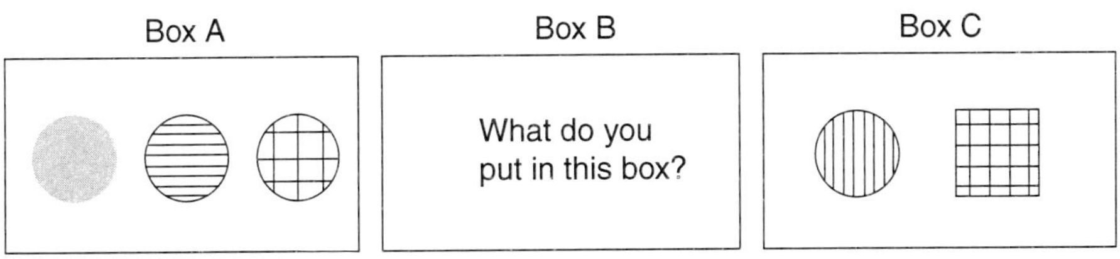

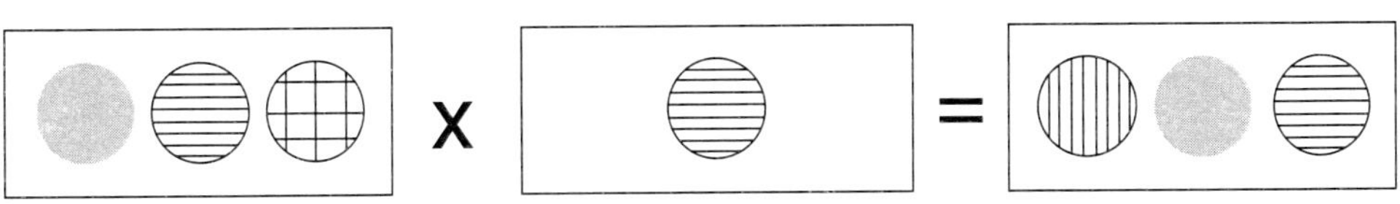

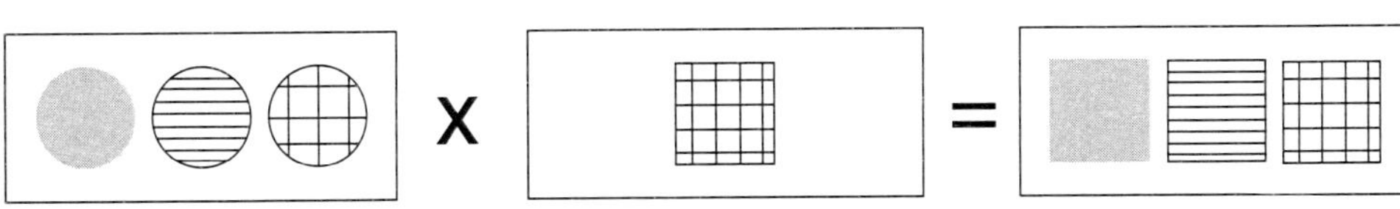

Now adding the two partial product piles and taking away a grey zero pair and a level zero pair we obtain:

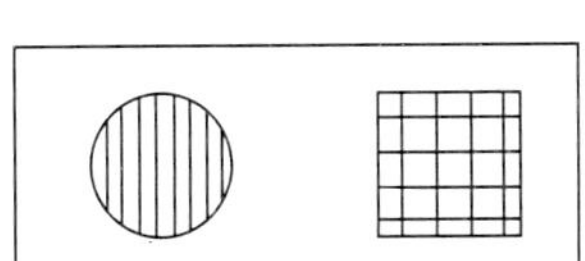

The first example is a way of saying the following:

$$(x^2 - 2x + 1)\ (x - 1) = x^3 - 3x^2 + 3x - 1$$

The second example tells you that:

$$(x^2 + x + 1)\ (x - 1) = x^3 - 1$$

in a more playful manner than usual!

Here is one that might surprise you. Work this one out yourself!

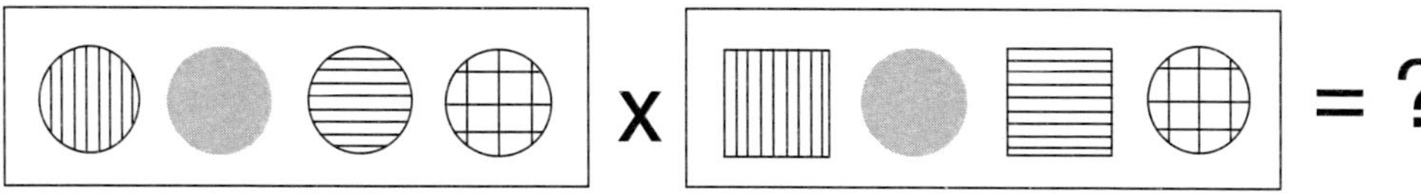

Does the above correspond to anything algebraic? If it does, what is it? If it does not, what spoils it?

The eight operators can be put into an arrow diagram as shown in Figure 41. This might make it easier on the memory to remember which operator does what!

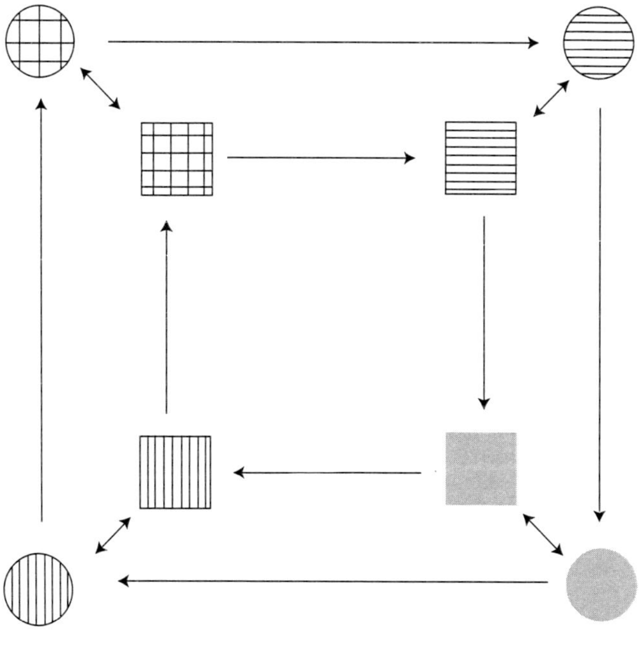

Figure 41

Each of the eight operators can be expressed as a sequence of one-way and two-way arrows. The level circle rule and the criss-cross square rule are expressed by means of just one arrow. The other operators will need two or three, one of them even needing four arrows.

It is left as an exercise for the reader to work out what each sequence is that expresses any particular operator. The expressions are not unique; each operator can be expressed by means of various different sequences of arrows.

SYSTEM 2

There is no reason why we should stay stuck with one lot of rules. Here are some suggestions for another set of rules for the same eight patterned counters:

Pattern changes:

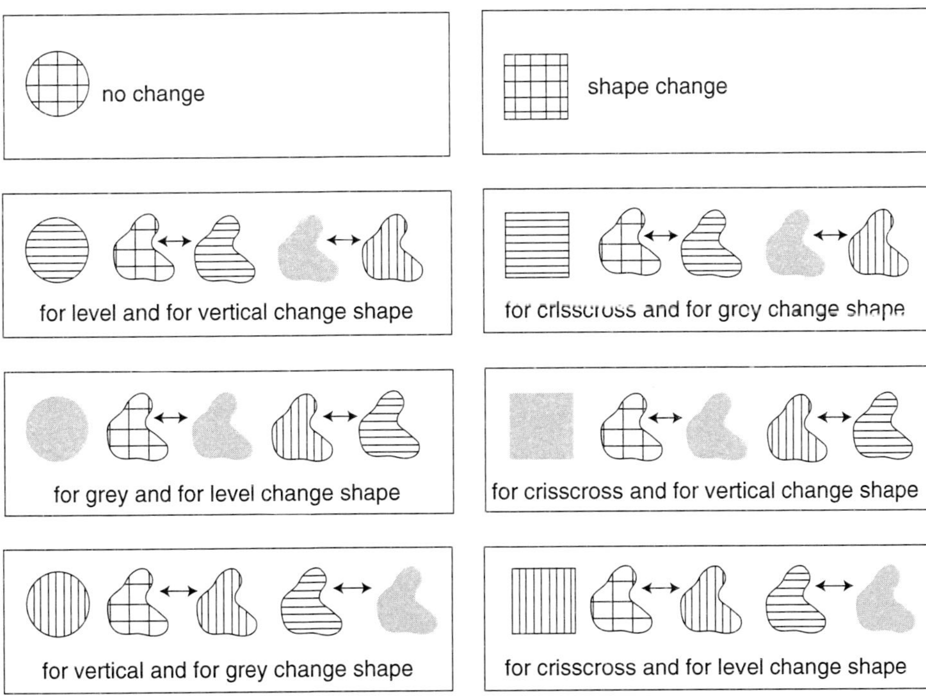

The above eight operators form what mathematicians call a group. This means, among other things, that applying any two of the operators one after the other can be replaced by applying just one of them. For example, it is easily checked that

applying 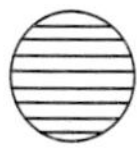 and then  can be replaced by 

Also, each operator has an inverse, in other words an operator that 'undoes' what the first operator has done.

Of course 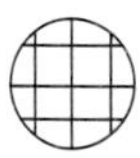 and 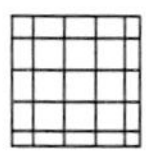 are their own inverses

But these are inverse pairs:

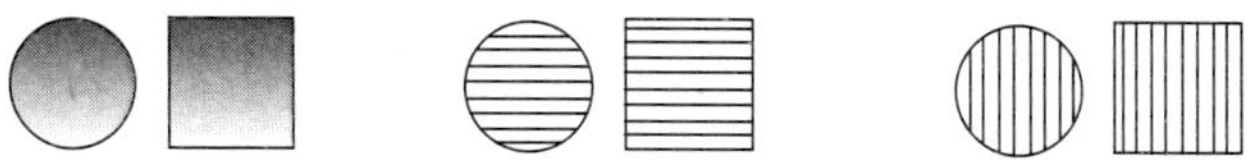

so doing one of them after the other comes to doing 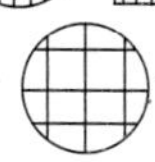 which is 'doing nothing'

It will also be checked that the associative law holds for the combination of these operators. When all these conditions are met, we say that the operators form a group.

It might be worth noting that combining any two of the above operations is always equivalent to just one of the above eight operations. There is an interesting cyclic order when you combine the upright, grey and the level circle operators. In fact, why do you not check that:

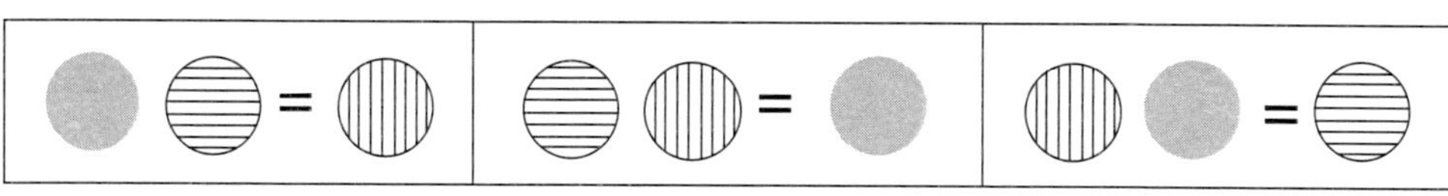

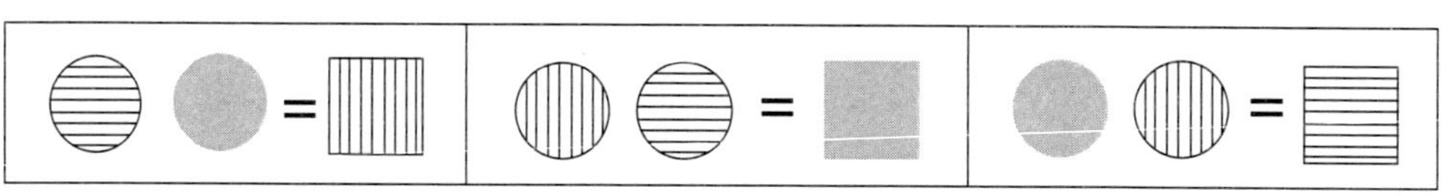

Another interesting consequence of this choice of the rules is that if we use any of the non-criss-cross operators twice in succession, all we get is a shape change, namely a change occasioned by the criss-cross square. Of course, if we use the same operator four times in succession, we are bound to return to our starting point, namely we have a sequence equivalent to the criss-cross circle.

Imagine a four-dimensional space, each pattern defining a dimension. Stepping away from the origin along one pattern direction, we increase the number of circles of that pattern. Stepping away from the origin in the opposite direction, we increase the number of squares of that same pattern. So we could regard circles as 'positive' and squares as 'negative'.

4. Symbolising

A pile of counters might be expressed symbolically as follows:

The number of criss-cross ones will be noted as positive or negative integers, depending on whether we are dealing with circles or squares. The level ones can be *i* numbers, the upright ones *j* numbers and the grey ones *k* numbers. So, for example, the pile

would be expressed as

$$2 + 3i - 2j - 4k$$

These are known as the quaternion integers. They were popular with mathematicians in the twenties, but the interest in them seems to have died out.

Figure 42 shows a good way of organising the eight elements of the quaternion game.

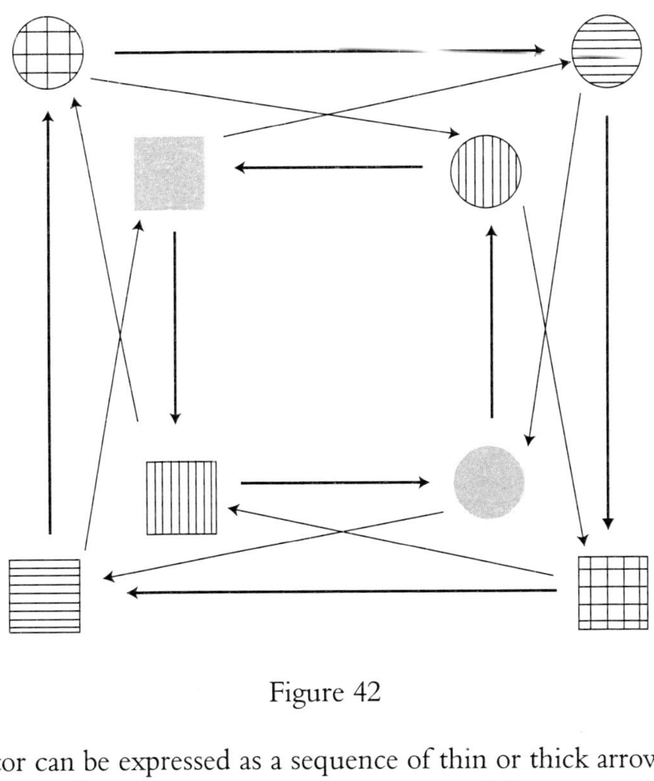

Figure 42

Each operator can be expressed as a sequence of thin or thick arrows. You will never need more than three arrows to express any of the operators. Some will need just one arrow, some will need two and some will need three. It is left as an exercise to work out how each operator is so expressed.

5. The Complex Integers

If we put aside the grey and the upright counters, then we are left with only criss-cross and level circles and criss-cross and level squares. The level circle will make us go round in a cycle of four one way, while the level square will make us go round in that cycle in the reverse order. The criss-cross square will take us halfway round the cycle.

If we play with only criss-cross and level counters, then we are playing at what mathematicians call the complex integers. So the complex integer game is a sub-game of the quaternion game.

We could equally choose the grey and the upright counters for embodying the complex integers. In this case the grey circle could be the no change and so the grey square would be the shape change. This would take us half way round the cycle. The upright circle would take us round one way, the upright square the other way round the cycle:

In fact, any non-criss-cross operator would give us a cycle of four, and the counters that we found along such a cycle would also do to play the complex integers game. Then, of course, if we only used the criss-cross counters, we would have an embodiment of the normal positive and negative integers with which all are familiar.

An interesting approach to studying the integers would be to play some quaternion games first, then reduce the number of dimensions to two and play the complex number game. After this we could reduce the number of dimensions to one and the study of positive and negative integers would seem like a piece of cake!

SOME NOTES ON FORMAL WRITING

The conventional notation for complex integers translates as follows from our circles and squares:

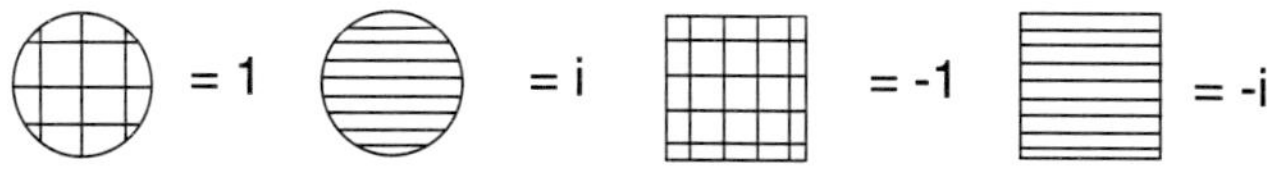

So for example

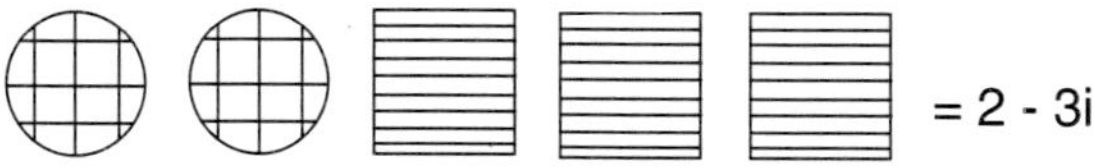

and since by our cyclic rules we have

we have

$i^2 = -1$

so i turns out to be the square root of –1.

So we can now solve equations such as:

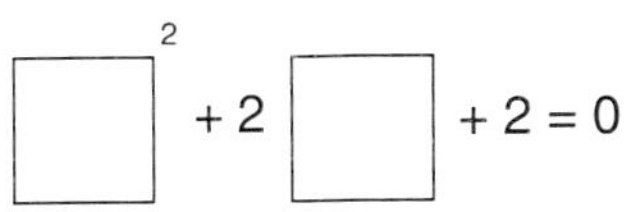

the solutions being:

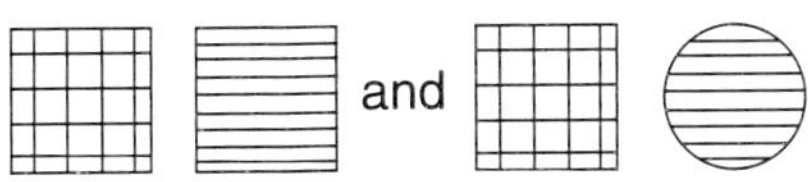

or formally:

$-1 - i$ $\qquad\qquad$ $-1 + i$

What happens if we use our first system and reduce it to two colours? In this case the pattern circle is just a level changer and so we have:

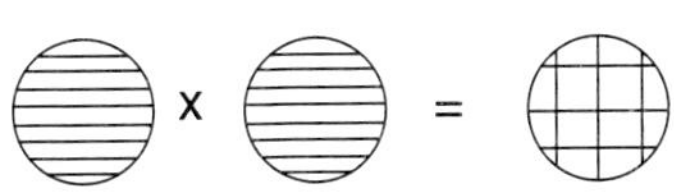

or $i^2 = 1$

and we have the embarrassment of finding non-zero elements whose product is zero, such as:

$(1 + i)(1 - i) = 1 - i^2 = 0$

a situation that cannot happen in the case of our second system reduced to two patterns.

Chapter VII

CYCLES OF THIRTY-ONE USING ATTRIBUTE BLOCKS

1. Introductory Remarks

In a standard set of attribute blocks (logic blocks) there are four attributes that can be varied: size, shape, colour and thickness.

There are four shapes: circle, square, triangle and rectangle (non-square). We can consider the square and rectangle as rectangular, while the circle and the triangle are non-rectangular. On the other hand we can consider the circle and the square as having a high number of symmetries (infinity and four, respectively), while we could think of the triangle and the rectangle as having a low number of symmetries (three and two respectively). So there will be two high symmetry (hysym) and two low symmetry (lowsym) shapes.

We shall also divide the attributes into strong ones and weak ones. Big, thick, red, hysym and rect (for rectangular) will be strong, while small, thin, blue, lowsym and non-rect will be considered weak.

The definition of the sum of two blocks is as follows:

The sum of two attributes of the same strength will be weak, but if one is strong and the other is weak, the sum will be strong.

For example a small thick red square, added to a big thin red circle will have to be thought of as small thick red hysym rect added to a big thin red hysym non-rect, which will result in a big thick blue lowsym rect block, namely in a big thick blue rectangle.

A block added to itself will yield the zero block, which is the small thin blue triangle, as this block has weak values for all its attributes.

2. The Construction of Sequences

Let us choose five non-yellow blocks but such that none of them can be obtained as the sum of any two of them or of any three of them or of the remaining four. For example, we could choose:

(i) big thick red square
(ii) big thick blue triangle
(iii) big thin blue triangle
(iv) small thick red triangle
(v) big thick red rectangle

Let us put them down in front of us in that order.

The next block will always be computed from the last five blocks by adding the first, the third, the fourth and the fifth blocks. It is not hard to work out that the sixth block will be a big thick red circle.

We could refer to each block by the strength of its five attributes. We could denote strong by 1 and weak by 0. We could write the strength values in the order size, thickness, colour, symmetry, rectangularity. Then the sequence of blocks that would be generated by our rules would be:

11111 11000 10000 01100 11101 11110 10111 00100 00001
01111 10100 01101 10010 01010 11010 10110 01011 10101
00010 00110 00111 01000 11100 10001 00011 01001 10011
00101 01110 11011 11001

and back to 11111.

If you look at the strength value of the same attribute in the above sequence of the blocks, you will see that they all follow the same cyclic pattern. This is the pattern:

1 1 1 1 1 0 1 1 1 0 0 0 1 0 1 0 1 1 0 1 0 0 0 0 1 1 0 0 1 0 0

Draw this sequence, twice over, on a long strip of paper, making sure the numerals are always at equal distances from each other on your strip. Then make four more identical strips. Then place the strips parallel to each other. You can slide your strips to the right and/or to the left, until you get them into a position in which there are no columns consisting of five 0s. You can then decide which strips represent which of the attributes, and construct the sequence of 31 blocks by following the columns of 0s and 1s from left to right (or the other way, if you prefer). There will be a way of arranging your strips so that you obtain the sequence above.

Here is the 31-block cycle shown with red as grey and blue as white (see Figure 43).

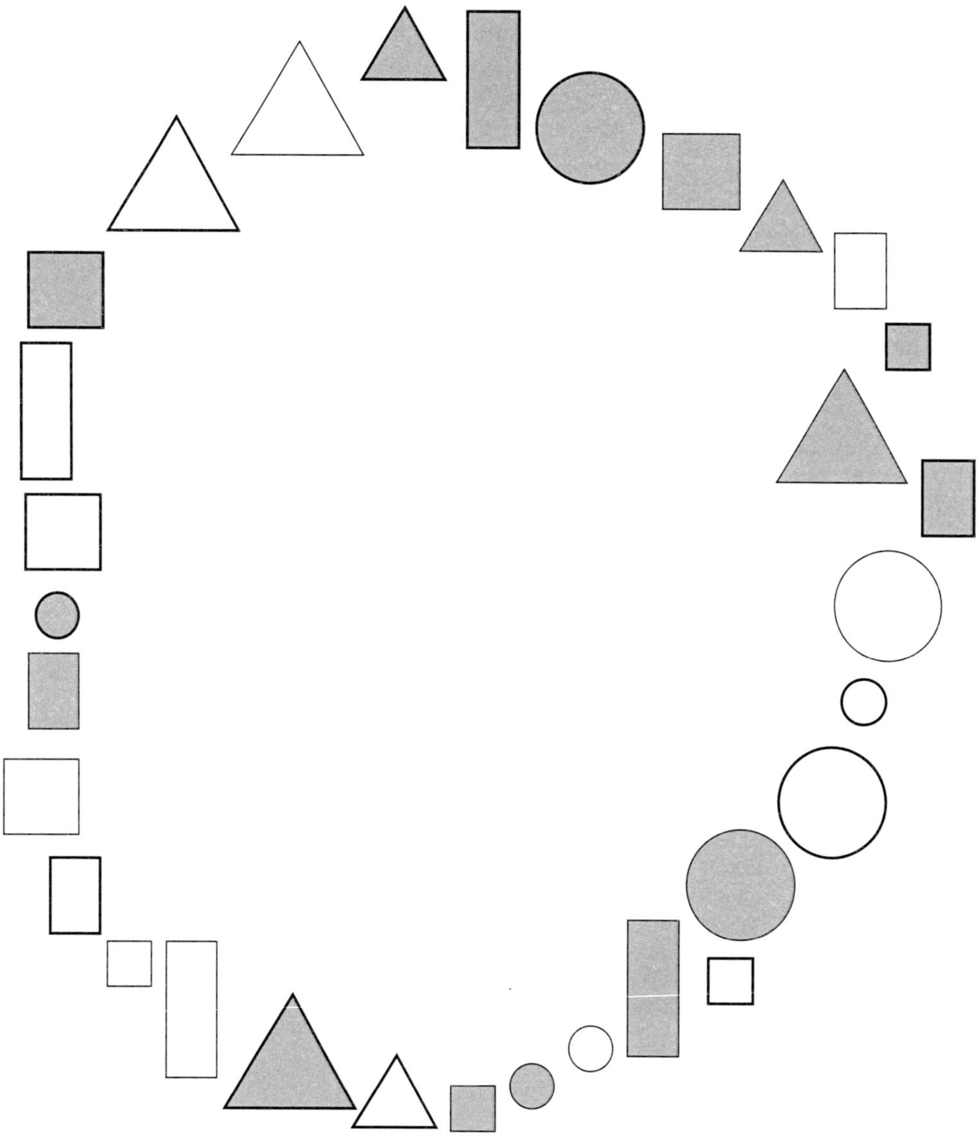

Figure 43

By the adding rule of the first, the third, the fourth and the fifth blocks in any run of five blocks, we can always construct the next block after the run of five being considered. But there are ways to tell what the next block is by looking only at one block.

It is easy to see that a thick block is always followed by a big block and a thin block by a small block. If a block has one or three of the properties big, thick, rectangular, then a thick block follows. If it has an even number, a thin block follows. If a block has an odd number of the properties big, thick, high symmetry, rectangular, then a red block follows. If an even number, then a blue block follows.

Circles and rectangles are followed by high symmetry blocks, squares and triangles by low symmetry blocks. If a block is a non-rectangular red or a rectangular blue, then a rectangular block follows; a rectangular red or a non-rectangular blue is followed by a non-rectangular block.

Figure 44 shows another 31 sequence of the same blocks.

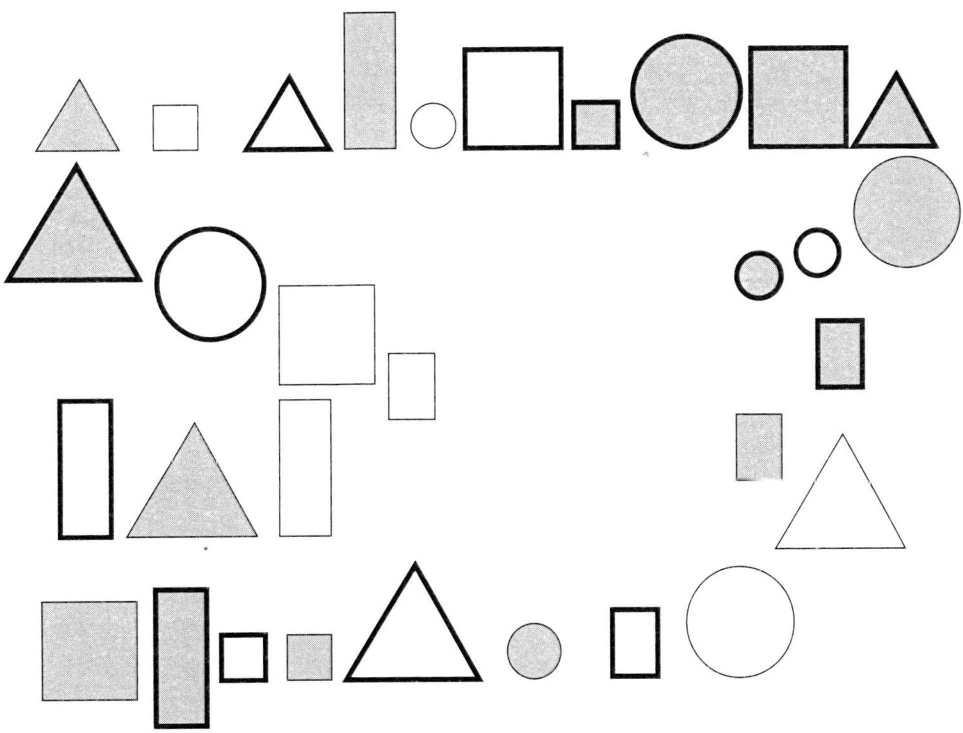

Figure 44

The rules of succession are quite different, and the order in which the strong and the weak attributes alternate is also different; it is this:

1 1 1 1 1 0 0 1 0 0 1 1 0 0 0 0 1 0 1 1 0 1 0 1 0 0 0 1 1 1 0

In fact, there are just two ways of ordering our 31 blocks, so that there are fixed succession rules for telling precisely which block comes after any given block. The two cycles given are examples of these ways.

You will find that in the above sequence the sum of the first, second, third and fourth blocks out of any run of five blocks, taken in the counterclockwise sense, will determine the block that follows the run of five selected.

The rules of succession for the above cycle are:

(i) An odd number out of the properties big, red, high symmetry, means that there is a red block to follow. If an even number of those properties is present, then a blue block follows.
(ii) An odd number out of the properties red, high symmetry and rectangular means that there is a big block to follow. If an even number of these properties is present, then a small block follows.
(iii) A red block is followed by a thick block and a blue block is followed by a thin block.
(iv) A thick block is followed by a high symmetry block, a thin block by a low symmetry block.
(v) A block having an odd number of properties big, thick, high symmetry or rectangularity means that there is a rectangular block to follow. In case there is an even number of these properties present, then a non-rectangular block follows.

3. Gathering up Subsystems of Fifteen Blocks

If you take any two blue blocks, their sum will always be another blue block. So the set of blue blocks forms a closed system. The same applies to the set of small blocks, to the set of thin blocks, to the set of low symmetry blocks and to the set of non-rectangular blocks. But there are lots more such subsystems.

All you have to do is to select one of the above subsystems, such as, for example, the set of blue blocks, and consider all the blocks that are the same number of blocks on from a blue block. For example, you could take all the blocks that are right next to a blue block, taking them either all clockwise, or all counterclockwise.

Let us go back to our first sequence. Each of these subsystems is such that there is a property which all its blocks have but none of the blocks outside it have. Such a property is called a characteristic property, as it characterises the blocks of a set.

The defining properties of our subsystems are listed below. Remember that 0 is an even number!

Characteristic properties of the 31 subsystems of 15 blocks:

(1) All the blue blocks.
(2) Blocks with an even number of high symmetry, big, thick, rectangular.
(3) Blocks with an even number of high symmetry, big, red, rectangular.
(4) Blocks with an even number of big, red, rectangular.
(5) Blocks with an even number of big, high symmetry, red.
(6) All the small blocks.
(7) All the thin blocks.
(8) Blocks with an even number of big, thick, rectangular.
(9) All the big red and all the small blue blocks.
(10) Blocks with an even number of big, high symmetry, rectangular.

(11) Blocks with an even number of red, high symmetry, thick.
(12) All the non-rectangular blocks (i.e. circles and triangles).
(13) All the red rectangular and all the blue non-rectangular blocks.
(14) Blocks with an even number of big, thick, red, high symmetry.
(15) All the thick rectangular and all the thin non-rectangular blocks.
(16) Blocks with an even number of red, thick, big.
(17) All the thick high symmetry and all the thin low symmetry blocks.
(18) Blocks with an even number of big, thick, high symmetry.
(19) All the big high symmetry and the small low symmetry blocks.
(20) Blocks with an even number of thick, high symmetry, rectangular.
(21) Blocks with an even number of any of the five strong attributes.
(22) All the thick red and all the thin blue blocks.
(23) All the low symmetry blocks.
(24) All the squares and all the triangles.
(25) All the red high symmetry and all the blue low symmetry blocks.
(26) All the big thick blocks and all the small thin blocks.
(27) All the big rectangular blocks and all the small non-rectangular blocks.
(28) Blocks with an even number of red, thick, rectangular.
(29) Blocks with an even number of: red, high symmetry, rectangular.
(30) Blocks with an even number of red, thick, high symmetry, rectangular.
(31) Blocks with an even number of red, big, thick, rectangular.

4. Reordering the Fifteen Blocks of a Subsystem

Any of the defined subsystems can be rearranged in such a way that all the strong and weak attributes should follow this cyclic pattern:

1 1 1 1 0 0 0 1 0 0 1 1 0 1 0

For example, let us try with the twenty-fourth set in our list, namely the set containing all the squares and all the triangles (see Figure 45).

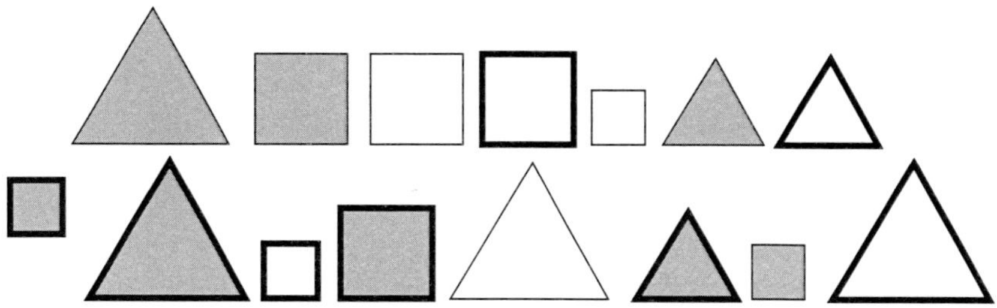

Figure 45

If you go round the above cycle in the clockwise sense, you will find all five attributes alternating between strong and weak as suggested above. You will notice that the high symmetry blocks always have the rectangularity property, and the low symmetry blocks always have the non-rectangularity property, so as far as symmetry and rectangularity are concerned, the above blocks are strong together as well as weak together, so these two properties really act as one property and not as two distinct ones.

The rules of succession for the set of squares and triangles are:

Big triangle or small square → red block next
Small thick or big thin → big block next
One or three of big, red, thick → thick block next
Big → square block next

5. Intersections of Systems of Fifteen

We can construct some easy intersections right away. We know, for example, that the blue blocks make up a subsystem of 15. If we look at the blue blocks in the above system of squares and triangles, we shall find exactly seven. In fact, the common part of any 2 of our 31 subsystems will yield a set of seven blocks. We shall call such a set a septet.

If you add any two blocks within a septet, you will get another block within that same septet. So a septet is also a closed system. In fact, having found the septet that is in two subsystems of fifteen, it will also be possible to find a third subsystem of fifteen in which our septet is included. So to characterise a septet, you might need to state the characteristic properties of all three systems of which the septet in question forms a part.

Taking our example of set number 24, intersecting it with set 1 (namely the set of all the blue ones), it is easy to check that set number 29 includes the blue ones that belong to set number 24, together with all the red blocks not in set number 24. So we can describe this septet by saying that every block in it is:

(i) blue
(ii) has zero or two of the properties rectangular and high symmetry
(iii) has zero or two of the properties red, rectangular, high symmetry

It is not hard to see that any two of the above properties imply the third one logically. A set of three such systems of fifteen will be called a triad of systems of fifteen.

What about three systems of fifteen that do not have a septet in common? You will find that in that case they will still have three blocks in common. What is more, the sum of any two of them is always the third block. We shall call such a set of three blocks a triad of blocks. What do four sets of fifteen have in common?

6. Different Ways of Setting Out Sets of Fifteen and Septets

Any set of fifteen blocks from our list of 31 sets can be set up in the form of a rectangle 3 × 5, with five columns of three blocks, but in such a way that each column is a triad of blocks, and the first three as well as the last three blocks in any row also form triads. Figure 48 shows a solution for our set of squares and triangles.

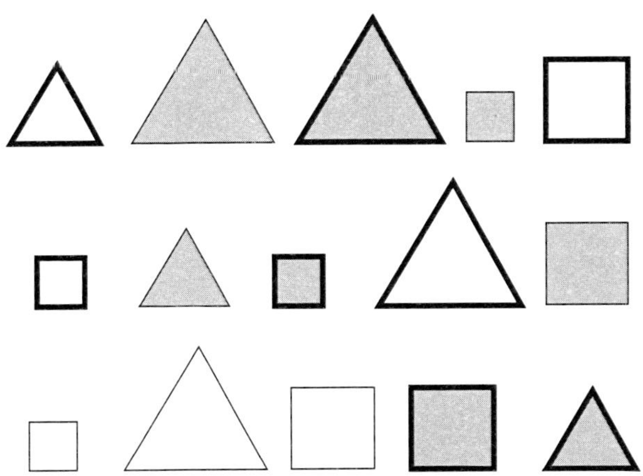

Figure 46

Or alternatively Figure 47 shows an arrangement of the set number 29.

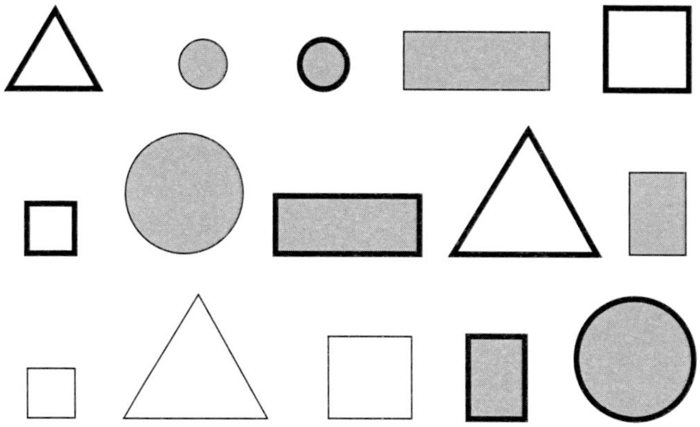

Figure 47

The septet which is common to the two sets of fifteen above is the set of blue blocks each set contains. We can place those in a triangular form to show the seven triads of blocks the septet contains. The arrangement is shown in Figure 48.

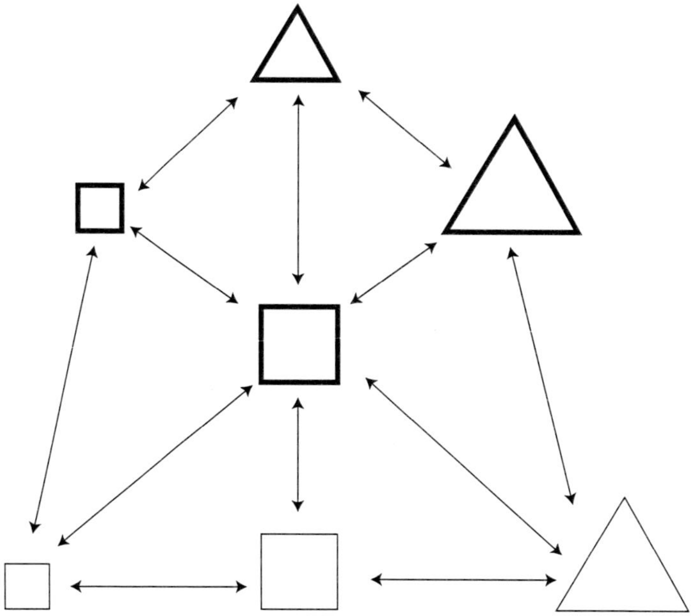

Figure 48

If you add any two blocks linked by an arrow, the sum will be the block which completes the straight line on which your chosen blocks are situated. If, however, you choose your blocks from the mid-points of two of the sides of the triangle, you will find the sum block at the mid-point of the third side. So the seven triads are made clearly visible in the diagram.

7. The Five-Way Symmetries of the 31 Blocks

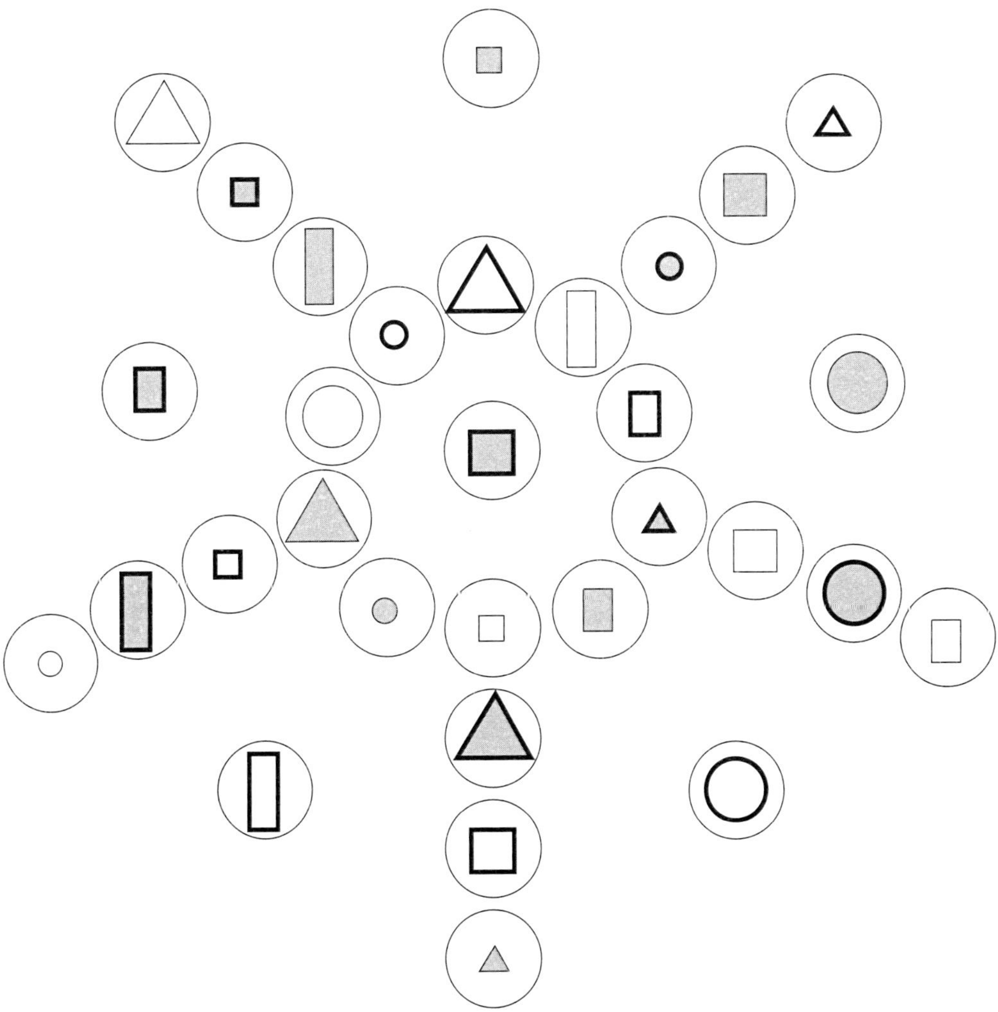

Figure 49

You should check that the blocks shown in Figure 49 on each of the five axes of symmetry form a septet. So each of these is included in three sets of fifteen.

The blocks in the septet lying in the axis next clockwise to the vertical axis belong to the sets of fifteen with the following characteristic properties:

(i) both or neither of high symmetry and redness
(ii) both or neither of bigness and rectangularity
(iii) an even number of strong non-thickness attributes

It is left as an amusement for the reader to find the characteristic properties of the sets of fifteen that contain the septets situated in the other axes of symmetry of our figure.

Each axis is a line of symmetry of the blocks if just one of the attributes is considered. For example, the vertical axis can be thought to be the colour axis, since it mirrors red ones into red ones and blue ones into blue ones. What is more, each block is reflected in this axis in a block that has the same number of strong non-red attributes. So the set of fifteen whose blocks have an even number of strong non-red attributes is symmetrical about this axis.

If we proceed in the clockwise sense to the next axis, we shall find that it is the thickness axis. Thick ones are reflected into thick ones and thin ones into thin ones in this axis. Reflecting any block in this axis, you will get a block with the same number of non-thick strong attributes. So the set of fifteen whose blocks have an even number of non-thickness attributes is symmetrical about this axis.

Let us move on to the next axis in the clockwise sense. This seems to be the symmetry axis. If we reflect in this axis, the strength of the symmetry of the reflected block is the same as the strength of symmetry of the one we are reflecting. Strong reflects into strong and weak into weak. So the set of fifteen whose blocks have an even number of non-symmetry attributes is symmetrically placed with regard to this axis.

The next axis, going round in the clockwise sense, will be seen to be the rectangularity axis. Rectangular blocks are reflected into rectangular ones, non-rectangular blocks into non-rectangular blocks. So the set of fifteen whose blocks have an even number of nonrectangular attributes is symmetrical about this axis.

The next axis is clearly the size axis. Big ones reflect into big ones and small ones into small ones in this axis. So the set of fifteen whose blocks have an even number of non-size attributes is symmetrically placed about this axis.

There is one set of fifteen which is so centrally placed that it is symmetrical about all five axes. The characteristic property of this set of fifteen is that of having an even number of the five strong properties. The ten blocks placed in the inner circle belong to this set, the other five being in the rays of four blocks, each time the third block out from the centre.

8. Duality Considerations

Since there are 31 sets of fifteen and 31 blocks, one might think that it would be possible to set up an interesting correspondence between the sets of fifteen and the blocks.

Would it be possible to find a correspondence in which to every three sets of fifteen containing one and the same septet there would correspond three blocks that formed a triad of blocks? If this were possible, then the triad of blocks so obtained could be made to correspond to the septet common to the three sets of fifteen. So blocks and sets of fifteen would correspond to each other, while septets and triads would correspond to each other.

Each set of fifteen is characterised by a property, defined by a certain number of strong attributes, of which the members of the set must possess an even number (including zero)! For example, the set of blue ones is defined by having zero red blocks. Then there are those determined by having either both or neither of two strong attributes. For example, the set of red thick blocks, together with the set of blue thin blocks, would constitute a set of fifteen. Or we could think of three or four or all five strong attributes; each time, if we select all the blocks with an even number of such attributes, we define a set of fifteen.

Let us take the following three sets of fifteen as our first example:

(i) zero red blocks (all the blue ones)
(ii) zero big blocks (all the small ones)
(iii) blocks with an even number of the attributes red and big (the big red ones together with the small blue ones)

Let us associate to each set the block which has those and only those strong attributes that define the corresponding set. This means that we have, respectively, these blocks corresponding to our sets:

(i) the small thin red triangle
(ii) the big thin blue triangle
(iii) the big thin red triangle

It is easy to check that any two of the above blocks will have the third block for their sum, and so they make a triad of blocks, corresponding to the triad of sets of fifteen.

Let us just look at one more example. Consider the following three sets of fifteen:

(i) blocks having an even number of the attributes: big, red, thick

(ii) blocks having an even number of the attributes: red, thick, rectangular

(iii) blocks having an even number of the attributes: big, rectangular

The corresponding blocks should be:

(i) the big red thick triangle
(ii) the small red thick rectangle
(iii) the big blue thin rectangle

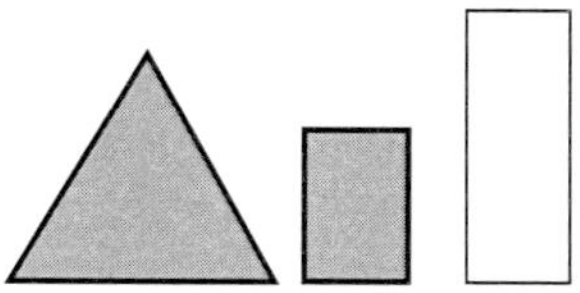

These three blocks do form a triad of blocks, whereas the three sets of blocks listed previously above have a septet in common:

(i) big thick blue square
(ii) big thick blue rectangle
(iii) small thin blue circle
(iv) big thin red square
(v) small thick red circle
(vi) small thick red triangle
(vii) big thin red rectangle

which can be placed in a triangle as shown in Figure 50.

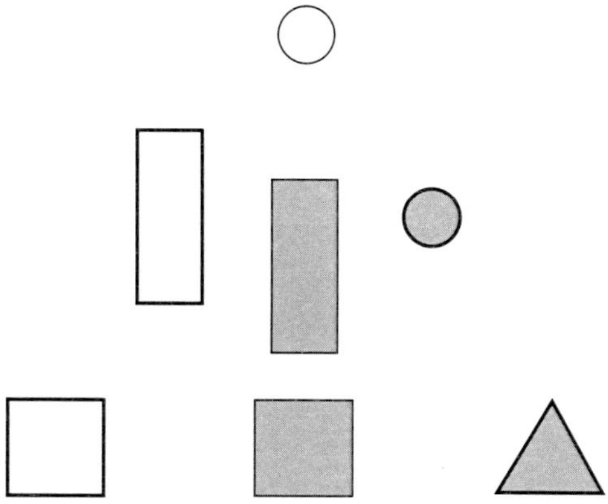

Figure 50

This triangle shows the seven triads contained in it.

8. Mathematical Notes for Those Interested

Mathematically speaking the set of 32 blocks has been considered as a concrete embodiment of the five-dimensional vector space, whose basis field is the two-element field. The two-element field has naturally two elements: the additive neutral or zero, and one other element. This field is embodied as the small blue thin triangle as the zero vector, together with any one of the other blocks as the second vector in the field.

In the games or puzzles, the zero vector has consistently been left out, thus making life a lot simpler. So mathematically speaking a triad, being a two-dimensional subspace, should have four elements, namely the 3 vectors (blocks) making up the triad together with the zero vector (the small thin blue triangle).

Similarly a septet, which is a three-dimensional subspace, actually has 8 vectors in it, adding the zero vector to the 7 vectors in the septet. A set of fifteen, which is a four-dimensional subspace (which I have been tempted to call a *quindecet*, but I in the end chickened out of such Latinate terminology), is also really a set of sixteen with the zero vector, not to mention the whole five-dimensional space, which has 32 vectors (blocks) in it.

The rules for generating the next vector in a sequence constitute what in mathematics is known as a matrix. For constructing a 31 cycle, you need a 5 × 5 matrix, containing only 0s and 1s. If we want to generate a 15 cycle, we need to work within a four-dimensional subspace. The sequence given on the second page of this chapter is generated by the following matrix:

0 1 0 0 0
1 1 0 0 1
1 1 0 1 1
0 0 0 1 1
0 0 1 0 1

The numbers in the first column are the multipliers of the size number of an input block, those in the second column are the multipliers of its thickness number, those in the third column operate on its colour number, those in the fourth column operate on its symmetry number and those in the last column operate on its rectangularity number. Adding these products obtained in the first row (using mod 2 addition, of course), we obtain the size number of the output block. The second row will provide the thickness number of the output block, the third row the colour number, the fourth row the symmetry number and the last row the rectangularity number of the output block.

For example, if we take the block (vector) 1 1 0 1 0 which is the big thick blue high symmetry non-rectangular block, in other words the big thick blue circle, then we shall have:

$(1 \times 0) + (1 \times 1) + (0 \times 0) + (1 \times 0) + (0 \times 0) = 1$ so the next one is big
$(1 \times 1) + (1 \times 1) + (0 \times 0) + (1 \times 0) + (0 \times 1) = 0$ so the next one is thin
$(1 \times 1) + (1 \times 1) + (0 \times 0) + (1 \times 1) + (0 \times 1) = 1$ so the next one is red
$(1 \times 0) + (1 \times 0) + (0 \times 0) + (1 \times 1) + (0 \times 1) = 1$ so the next sym is high
$(1 \times 0) + (1 \times 0) + (0 \times 1) + (1 \times 0) + (0 \times 1) = 0$ so the next is non-rect

So the next block after the big thick blue circle must be a big thin red circle, a circle being a high symmetry and non-rectangular block.

As mentioned previously, there are two ways in which the strong and the weak values of the attributes can alternate in a 31 cycle (not counting reversing the order as different). It is possible to prove this, but such proof goes beyond the elementary, so it is not included here. As far as I am aware, such proof was first found by Dr R O Davies, of Leicester University, UK, who would, I am sure, be glad to furnish details to any interested enquirers.

For the four-dimensional case there is only one way in which the strengths of the attributes can alternate (if we do not count the reverse order as another order). This is also the case for the three-dimensional case, in which case the cyclic order will always be 1 1 1 0 1 0 0 and back to 1 or the reverse. All septets should be able to be arranged in that cyclic order.

Chapter VIII

A SEQUENCE OF THIRTY-ONE

1. A Sequence of 31 Determined by Some Rules

You might have noticed that the *f*'s, the *i*'s, the *t*'s and the *s*'s occur as shown by the letter Y (for Yes) as opposed to not occurring as shown by the letter N (for No) expressed by the following cyclic sequence:

Y Y Y N Y Y Y Y Y N N Y N N Y Y N N N N Y N Y Y N Y N Y N N N

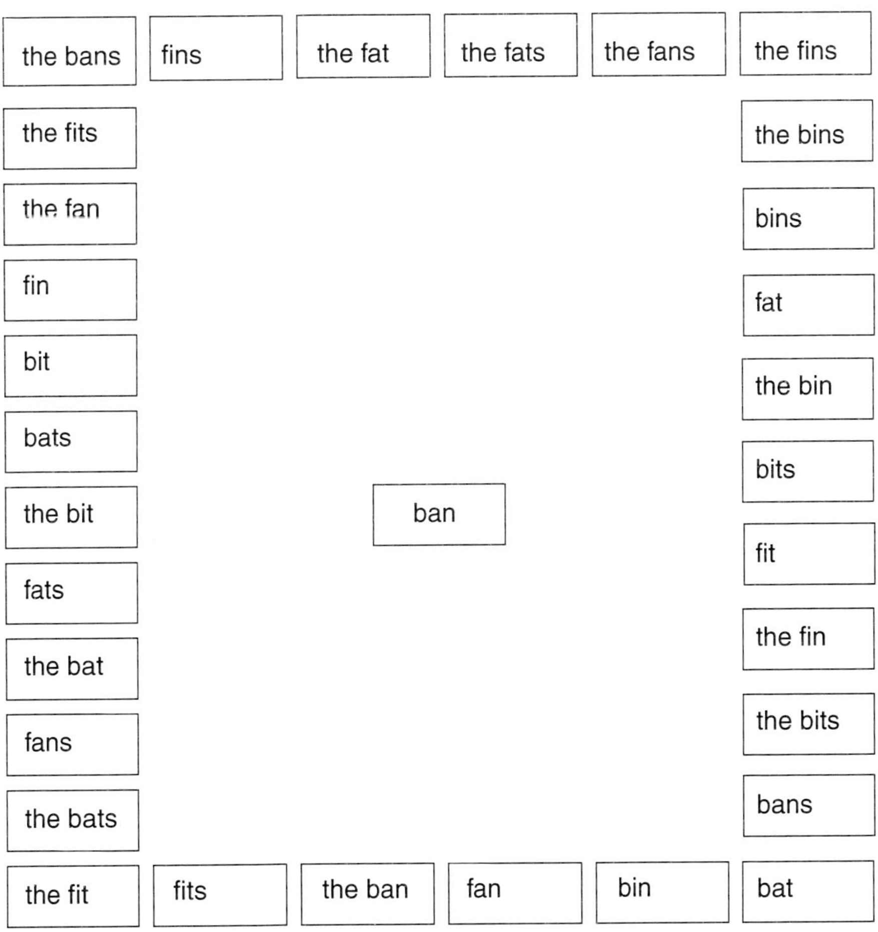

A song could be learned where Y is replaced by the corresponding note being higher than the previous note, and N is replaced by the corresponding note being lower than the previous note. Here is a pentatonic suggestion (a waltz?), the upper notes being shown by a star:

la do re / do re mi / sol la★ do★ / la★ sol★ la★ / sol★ mi sol★ /
la★ sol★ mi / re do re / mi re mi / re mi re /do la do /
re

The clockwise rules for getting the next word are the following:

f → *the* next
b → no *the* next
the with *a* or no *the* with *i* → *f* next
the with *i* or no *the* with *a* → *b* next
the with *n* or no *the* with *t* → *i* next
the with *t* or no *the* with *n* → *a* next
the or *s* but not both → *t* next
Both or neither of *the* or *s* → *n* next
the → *s* next
No *the* → no *s* next

Knowing these rules, it is possible to write down the whole sequence. It is somewhat more difficult to guess the rules, given the sequence!

To check the degree of difficulty in discerning rules, why not try to see what the rules would be if you skipped a word all the time? You would still get through all the words as 2 is not a factor of 31. In fact, you could skip any fixed number of words, as long as you kept skipping the same number, and you would still bring in all 31 words, and there would be some rules which would generate the sequence so obtained. This is so because 31 is a prime number.

If you prefer to work with numbers, the same things can be done with the factors of 2310. We have $2310 = 2 \times 3 \times 5 \times 7 \times 11$, and so the *the*-ness, and the presence of the letters *f*, *i*, *t* and *s* are simply replaced by the number in question having the factors 2, 3, 5, 7 or 11 respectively.

Of course, you could have pictures with five different things which are either in a picture or they are not, for example a boy, a girl, a dog, a cat and a house, or anything else anyone would like to think of. Or you could make little towers with coloured cubes. You would need five colours and sixteen cubes of each colour for making all the 31 little towers.

Table 9 shows the rules for the next word but one.

Table 9

Word you are looking at	Next word but one
The with an *a* word, or no *the* with an *i* word	→ *the*
The with an *i* word, or no *the* with an *a* word	→ no *the*
One or three of *the*, *f* or *t*	→ *f*
None or two of *the*, *f* or *t*	→ *b*
One or three of *the*, *f* or *s*	→ *i*
None or two of *the*, *f* or *s*	→ *a*
The with a *b* word or no *the* with an *f* word	→ *t*
The with an *f* word or no *the* with a *b* word	→ *n*
f word	→ *s*
b word	→ no *s*

Here are some ways of selecting sequences of fifteen out of our sequence. These should be distributed as the Ns are distributed among the Ys in the pattern given before. The easiest ones to find are:

(1) the non-*the* words
(2) the *b* words
(3) the *a* words
(4) the *n* words
(5) the non-*s* words

There are ten easy ones to find when we use just two letters. These are:

(6) *ba* words and *fi* words
(7) *an* words and *it* words
(8) *bn* words and *ft* words
(9) *ts* words and *n* no *s* words
(10) no *the b* words and *the f* words
(11) no *the a* words and *the i* words
(12) *a* no *s* words and *i s* words
(13) no *the n* words and *the t* words
(14) *b* no *s* words and *f s* words
(15) no *the* no *s* and *the s* words

There are ten more sets in which we consider three of the letters, five more in which we consider four of the letters and just one in which we consider all the five letters (I am counting *the* as a letter).

Take, for example, the three letters following *the*. We must be looking for the appearance of fin, fat, ban, bit. Words containing these will fit the No pattern of our sequence of Ys and Ns.

In each case we have to consider those words in which the letters to be considered out of 'the fits' number either zero or two or four. This gives all the 31 ways.

Having selected a Yes–No split, we can rearrange the words in the No part of the split, so that all the letters in 'the fits' should conform to the following pattern:

Yes Yes Yes Yes No No No Yes No No Yes Yes No Yes No

For example, we could choose to work on the words that have fin, fat, ban, bit in them. These words are obtained by choosing the part 'fit' and finding the words which have either two or none of these three letters in them. Here are the fifteen words in the required order:

the bits, the fats, fins, the fins, fin, the bit, the ban, the bans, the fat, bit, bans, fats, the fin, bits, fat

and it is easily checked that all the letters *the*, *f*, *i*, *t* and *s* satisfy the requirement of the Yeses in the series:

Yes Yes Yes Yes No No No Yes No No Yes Yes No Yes No

It makes an amusing afternoon's entertainment for guests who want to be challenged to work out the order for any of the other 30 sets of fifteen that can be constructed from our 31 words!

Here is one set. Check whether I have placed the words in the correct order and find out which letters out of 'the fits' I have used to make the set.

fans, the fit, the fats, the fins, the bits, bins, bit, fat, the bat, bats, fin, the fan, the bans, fits, the bin

2. The Use of Coloured or Patterned Cubes

Of course it is much easier to do all this with patterned or coloured blocks. Figure 51 shows a corresponding sequence using patterned cubes.

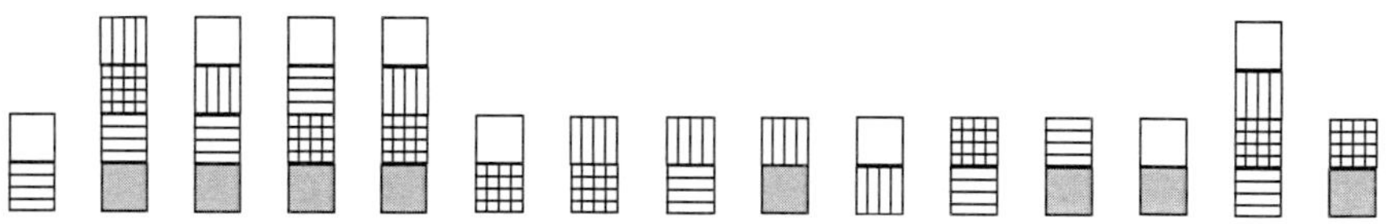

Figure 51

where grey means *the*, level means *f*, criss-cross means *i*, upright means *t* and white means *s*, the lack of the above patterns meaning no *the*, *b*, *a*, *n* and no *s* respectively.

The puzzle, in its general form, could be set up as follows:

(i) Collect forty cubes, eight cubes of each of the above five patterns.
(ii) Choose a number between 1 and 5 inclusive. This is your chosen number.
(iii) Out of the five patterns choose as many patterns as your chosen number.
(iv) Using either 32 or all your cubes, make fifteen little towers, never repeating a pattern in the same tower, but so that in each tower there should be an even number of cubes of your chosen patterns. Remember that zero is an even number.
(v) Arrange your little towers in a circular pattern so that each one of the five patterns shall occur according to the rule

Yes Yes Yes Yes No No No Yes No No Yes Yes No Yes No

It will always be possible to make the above arrangement with any choice of the number of patterns being considered.

You might be wondering what happens if you choose 1 as your chosen number. That means you choose one pattern, say grey. You have to make all the towers with an even number of grey ones, and never put more than one of any pattern into a tower, including grey ones. This is where it is good to remember that zero is an even number. So we make all the little towers without any grey ones in them.

Another interesting problem is to find an arrangement of all the 31 little towers so that all the 31 sets of 15 are symmetrically placed. This can be done by placing the towers in a pentagonal pattern.

Figure 53 shows a suggested pattern in which the sets of fifteen are all symmetrically placed. I think this is true of the sets of seven, but the sets of three (the triads) have sometimes had to be content with rotational symmetry.

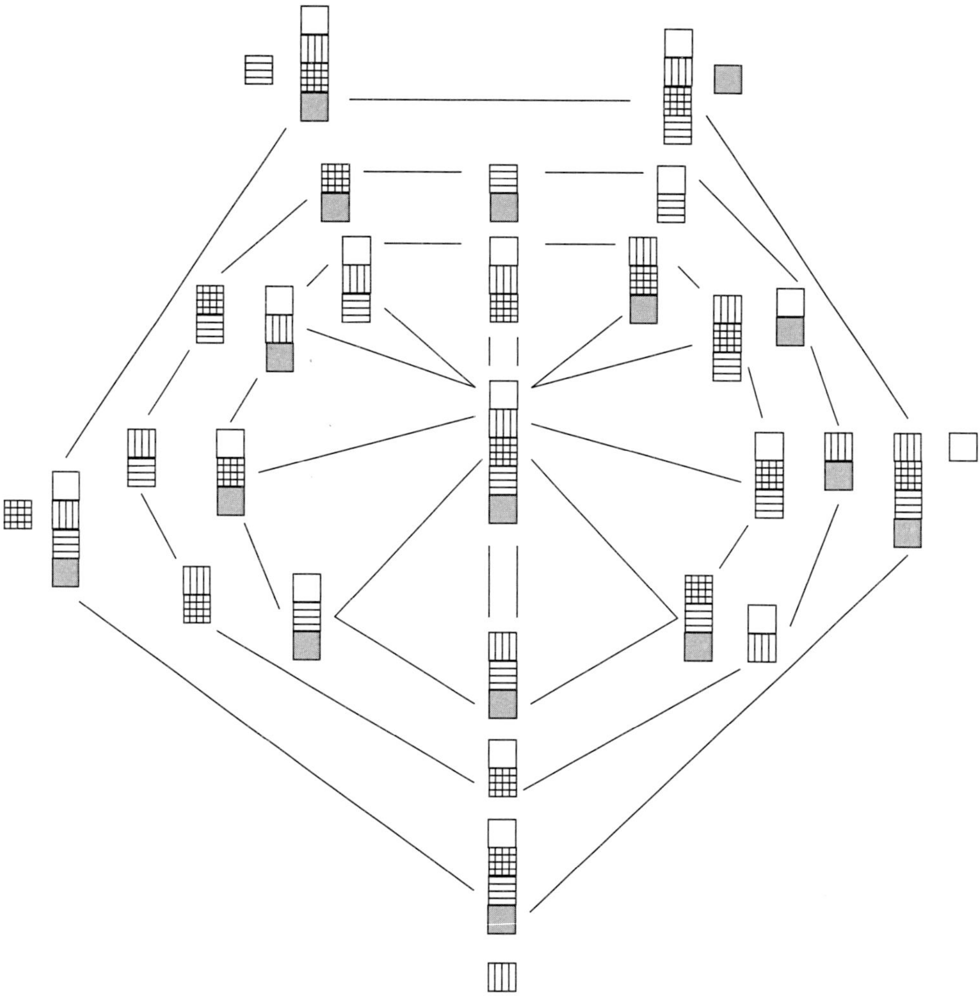

Figure 52

Figure 53 shows some sets of fifteen in the correct order.

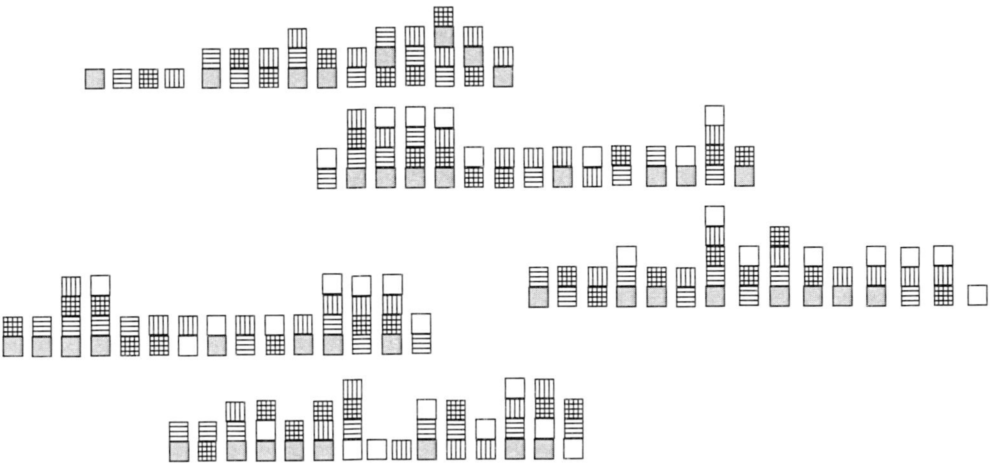

Figure 53

It is unfair to know a little algebra. Some people might spend hours fixing the Yes–No order for any particular set of 15 taken out of a set of 31, but if you happen to know that it is enough to start with five towers in which the total number of any of the colours is either zero or two or four in the first, second and fifth towers, then you will be much quicker than your friends. You will be even quicker if you make sure you choose the sixth tower, so that there is an even number of each colour in the second, the third and the sixth tower, followed by choosing the seventh tower so there will be an even number of each colour in the third, the fourth and the seventh tower and so on.

If you carry on like this, you will miraculously have made sure that all the colours come out according to the required Yes–No rule.

In the case of the puzzle with the 31 words, it is somewhat more difficult to formulate the corresponding rule, so you can amaze your friends even more by translating the five colour problem into the corresponding five parts of a word problem!

It might be even better to have a 31 sentence puzzle. Just think of two adjectives (or pronouns), two nouns, two verbs, two temporal complements and two spatial complements. You can make 32 sentences by choosing differently 32 times. Then appoint one of the sentences as the zero sentence, and you can play the same games as before with the remaining 31 sentences.

You could select your sentences from Table 10.

Table 10

These	girls	are playing	all the time	in the house
Those	boys	are working	every morning	in the fields

The sentence in the first row could be appointed the zero sentence. You could then define a set of fifteen, for example by choosing all the sentences in which two or none of the parts *those*, *boys* and *in the fields* occur. You would get the following:

Those boys are playing all the time in the house.
These boys are playing all the time in the fields.
These girls are working all the time in the house.
These boys are working every morning in the fields.
Those girls are playing all the time in the fields.
These boys are working all the time in the fields.
These boys are playing every morning in the fields.
Those boys are working every morning in the house.
Those boys are working all the time in the house.
These girls are working every morning in the house.
Those girls are working all the time in the fields.
These girls are playing every morning in the house.
Those boys are playing every morning in the house.
Those girls are playing every morning in the fields.
Those girls are working every morning in the fields.

Chapter IX
THE THIRTEEN GAME

1. Introducing the Thirteen Game (Shapes and Towers)

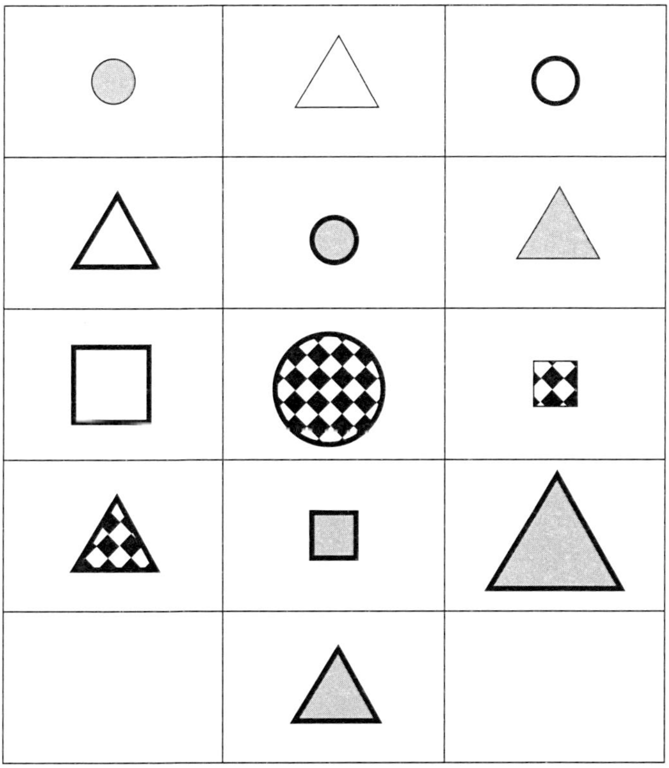

Figure 54

There are thirteen ways in which it will be permitted to pick out sets of four shapes out of the thirteen shapes shown in Figure 54. They are:

(1) Take all the white ones.
(2) Take all the circles.
(3) Take all the unframed ones.
(4) Take the small framed triangles and the unframed circle.
(5) Take the small framed grey ones and the white unframed one.
(6) Take all the grey triangles and the white circle.

Before I can show you the other seven ways, I shall have to split the properties of the shapes chosen into ones with low value, ones with medium value and ones with high value (see Table 11).

Table 11

	Dimensions	Shading	Shape
High values	Big framed	Spotted	Square
Medium values	Small framed	Grey	Triangle
Low values	Small unframed	White	Circle

Pick two of the three attributes out of:

(i) dimensions
(ii) shading
(iii) shape

Call them A and B.

Pick four shapes as follows:

(i) high value A with medium value B
(ii) medium value A with high value B
(iii) high value A with high value B
(iv) low value A with low value B

Since you can carry out the above in three ways, this gives ways (7), (8) and (9) for picking a set of four shapes.

Here are the remaining four ways of doing it (namely ways 10, 11, 12 and 13):

Pick one shape which has no low value attribute.
Look at two of the three attributes of this chosen shape.
If one of these is of medium value and the other of high value, then pick another shape in which both these attributes are of medium value and the third attribute of low value.
If both of the chosen attributes are of medium value, then pick another shape in which these same two attributes are of high value and the third one is of low value.

Here is an example of how to pick according to this last set of rules:

Pick this shape:

It has medium value dimensions, high value shading and medium value shape. Dimensions and shading are medium and high, so we pick a medium–medium

so it will have to be a low value for the third attribute. Medium value dimension is small framed, medium value shading is grey, and low value shape is a circle, so we pick:

Now take dimension and shape, which both have medium values, so we have to pick high values for these attributes and low for the third attribute. High values for dimension and shape are large and square, and for shading the low value is white, so we pick:

Now we take shading and shape, spotted being high and triangle being medium values; we must look for medium shading and medium shape and obviously low value dimension. Medium value shading is grey, medium value shape is triangle, and low value dimension is small. So we have to pick a small grey triangle, namely:

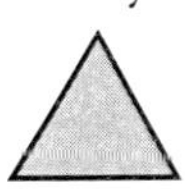

So our set of four shapes, using our rules, is:

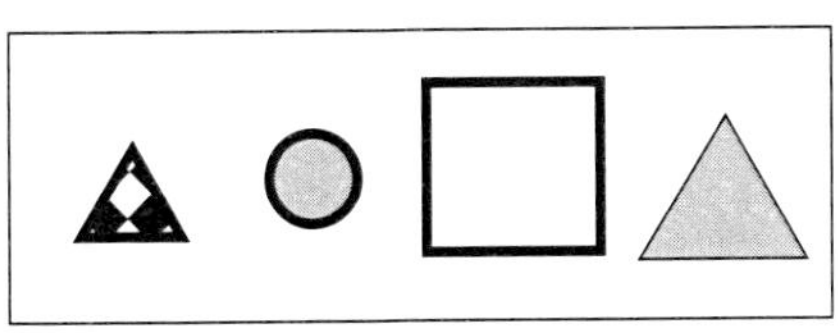

If you do not have a set of logic blocks (attribute blocks) to use as your shapes, you can use coloured cubes for making little towers, using three colours. You can choose your own colours; I shall indicate them as white, grey and spotted. Each tower will have three cubes in it: a bottom one, a middle one and a top one. This is how you can make them, given in the same order as the shapes. You can easily work out the relationship between the shapes and the towers, I could give the hint that white means low value, grey, medium value and spotted means high value (see Figure 55).

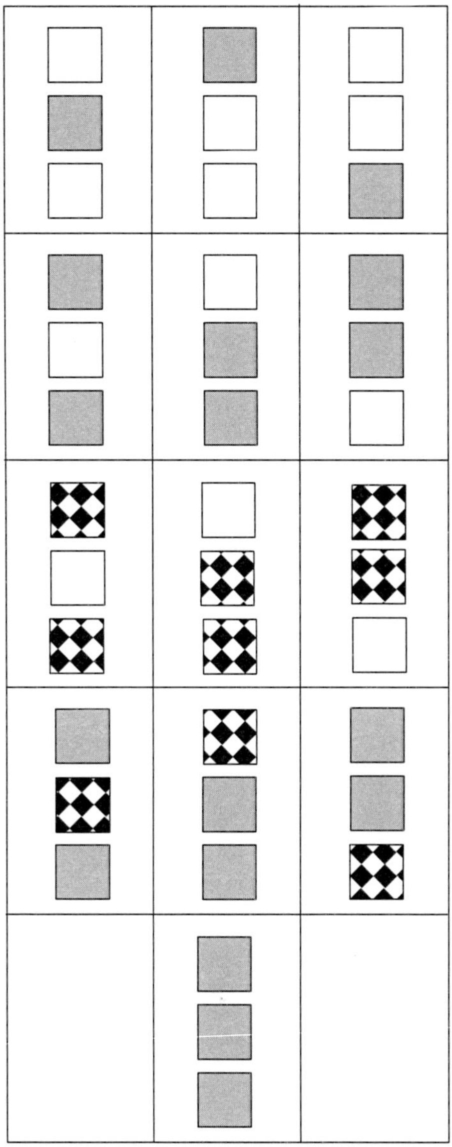

Figure 55

Each 'level' could correspond to an attribute, white could mean low value, grey medium value and spotted high value. The thirteen ways of making permitted sets of four can now be expressed as follows:

(1) The top one is white.
(2) The middle one is white.
(3) The bottom one is white.

(4), (5) and (6) Two of the levels are of equal shading but not spotted

(7), (8) and (9) On two of the levels there are either two white ones or a spotted with either another spotted one or a grey one.

(10), (11), (12) and (13) Pick one tower with no white and three towers with just one white in it, but such that:

(i) if in the no white tower two levels are grey, in a one white tower these two levels must be spotted (and, of course, the third one must be white).

(ii) if in the no white tower one level is grey and another is spotted, there must be a one white tower in which both these levels are grey (and the third level, of course, is white).

You could now pick any two towers (or any two corresponding shapes, for that matter!) and one and just one of these rules will determine the other two towers (or shapes) that you must have in your permitted set of four.

Another fun puzzle would be to place nine of your thirteen towers (or shapes) in a 3 × 3 arrangement, so that every row of three, as well as every column of three and the two diagonals, should contain towers (shapes) of a permitted set of four. The remaining four will also make a permitted set of four.

Figure 56 shows one possible solution.

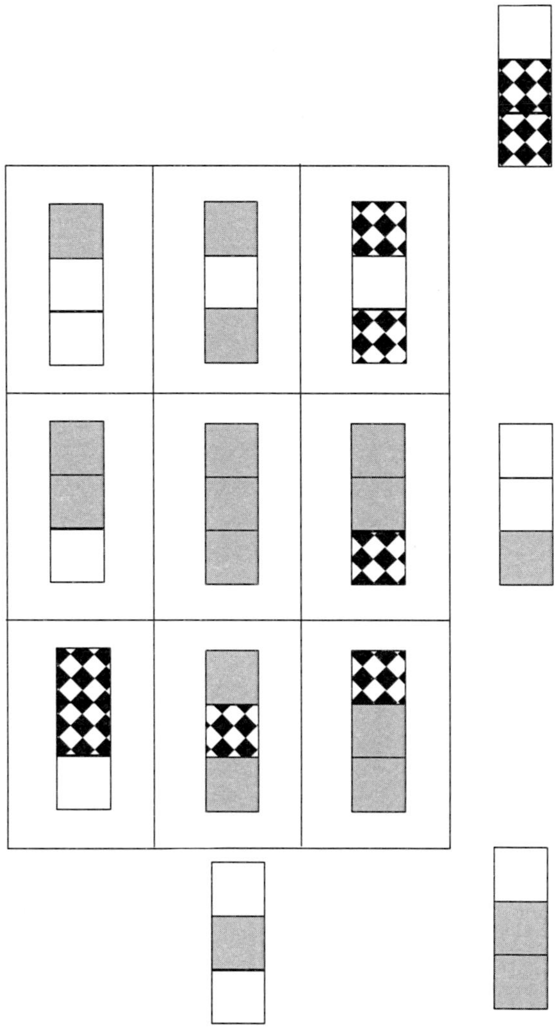

Figure 56

Another little problem you might like to solve is this one:

Arrange the thirteen towers in a circle, so that in any run of seven towers, the first, the second, the fifth and the seventh towers should form a permitted set of four. If you want to solve the problem yourself, don't look at the solution shown in Figure 57.

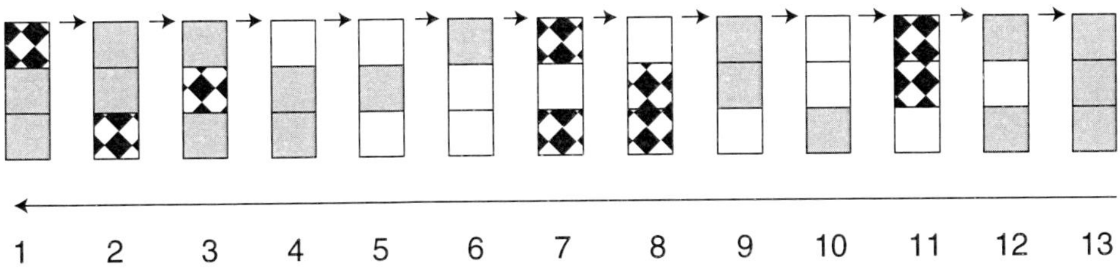

Figure 57

You could now play the following game with the above cycle:

You know that in every run of seven towers the first, the second, the fifth and the seventh form a permitted set of four towers. So you can pick any two towers and find where the other two towers are that make up the permitted set by applying that rule. You can just reason on the numbers given to the towers as suggested above.

For example, take towers 3 and 10. Where are the missing towers from the set?

{10, 11, 1, 3} form a permitted set of four, counting tower 10 as the first in the run of seven (10, 11, 12, 13, 1, 2, 3). We can see that one of rules (7), (8) or (9) can be applied to the towers in the set of four, as if we look at the upper and middle levels we can see that they are either both white or at least one of them is spotted, the other being either spotted or grey.

We can say the same thing of the upper and lower levels if we take the set {1, 2, 5, 7}. Finally we can say the same thing of the middle and lower levels if we take the set {2, 3, 6, 8}.

If we want to find the sets in which one level is always white, they are:

{4, 5, 8, 10}, {5, 6, 9, 11}, {6, 7, 10, 12}

in which the upper, then the lower, then the middle level respectively stay white.

The sets in which at two given levels the shading is equal, but never spotted, are {13, 1, 4, 6} for middle and lower levels, {9, 10, 13, 2} for middle and upper levels, and {12, 13, 3, 5} for the upper and lower levels.

The sets of four where we pick one without white and three with one white each are:

{7, 8, 11, 13}, {8, 9, 12, 1}, {11, 12, 2, 4} and {3, 4, 7, 9}

No doubt you will have noticed that any two sets of four have one and only one tower in common and that any given tower belongs to exactly four sets.

You can also use following the set of numbers for doing exactly the same things:

{2, 3, 5, 15, 10, 6, 225, 100, 36, 60, 90, 150, 30}

2. Links Between Numbers and Towers (the Cube)

The translation between little towers and numbers would go something like this:

(i) 2 but not 4 is a factor means middle level is grey
(ii) 3 but not 9 is a factor means upper level is grey
(iii) 5 but not 25 is a factor means lower level is grey
(iv) 2 is not a factor means middle level is white
(v) 3 is not a factor means upper level is white
(vi) 5 is not a factor means lower level is white
(vii) 4 is a factor means middle level is spotted
(viii) 9 is a factor means upper level is spotted
(ix) 25 is a factor means lower level is spotted

You can now work out the rules for selecting permitted sets of four numbers.

Then you can play the same games with the numbers that we have played with the shapes and with the little towers, putting nine of the numbers in a 3 × 3 square, and then putting all the numbers in a thirteen cycle so that in any run of seven numbers the first, the second, the fifth and the seventh numbers form a permitted set of numbers.

Here is one way of using the thirteen numbers in a more sophisticated way:

Make a pattern like the one below, write the numbers on it as shown and then cut it out and fold it together with the numbers on the outside to make a cube. You will be able to find the thirteen axes of rotation of the cube, given by the thirteen numbers. The sets of four numbers will give sets of four axes all in a plane, these being the planes of symmetry of the cube. There are four exceptions: in these cases only three of the axes are in the same plane (which is not a plane of symmetry); the fourth one is always perpendicular to that plane.

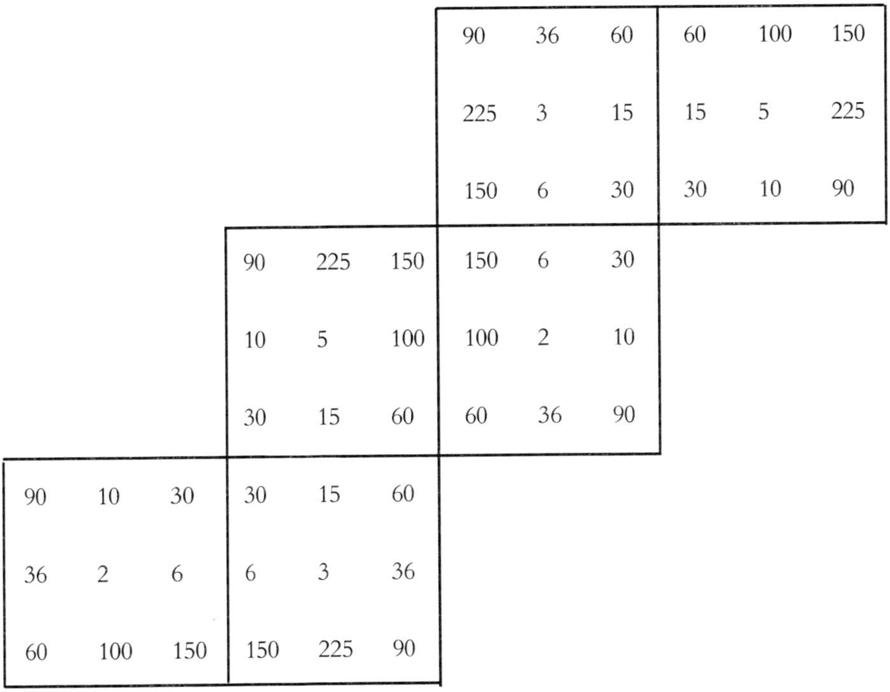

Figure 58

There are thirteen permitted sets of four, but there are also thirteen shapes, or towers or numbers. You might have wondered if there might not be some connection between these thirteens. As a matter of fact, there is. Table 12 shows a suggested relationship.

Table 12

3	5	15	225	<——>	2
2	5	10	100	<——>	3
2	3	6	36	<——>	5
2	15	30	60	<——>	225
3	10	30	90	<——>	100
5	6	30	150	<——>	36
2	225	90	225	<——>	15
3	60	150	100	<——>	10
5	60	90	36	<——>	6
60	225	6	10	<——>	60
90	100	6	15	<——>	90
150	36	10	15	<——>	150
30	225	100	36	<——>	30

If you have made your cube with the numbers on, you can check that, in the case of the first nine sets, the corresponding number on the right refers to an axis which is at right angles to the plane in which the four axes lie which are given by the four numbers of the permitted set.

In the case of the last four, you will notice that the corresponding number is one of the numbers of the permitted set. This axis is perpendicular to the plane in which the three axes lie that are given by the second, the third and fourth numbers of the permitted set. You will also find that this plane cuts your cube in the form of a regular hexagon.

You can also check the following:

Choose one of the thirteen numbers. Find the four sets to which this number belongs. Find the corresponding four numbers in the right-hand column. These four numbers should form one of the thirteen permitted sets. If this is not so, you have made some mistake somewhere.

So our innocent games with shapes and little towers, and finally with numbers, are really inspired by the axes of rotations of the cube, nine of the permitted sets of four leading to the nine planes of symmetry of the cube, the remaining four permitted sets of four leading to the four ways in which we can cut the cube in halves along a hexagonal cut.

3. Orthogonality

There is still more you can do with your thirteen shapes - towers - numbers. You can put them in pairs. But some pairs will be permitted and others not. The permitted pairs will be called ortho-pairs.

I shall distinguish three kinds of towers:

(i) with only one non-white level, these are the *one-ers*
(ii) with just two non-white levels, these are the *two-ers*
(iii) with no white levels at all, these are the *three-ers*

You can make an ortho-pair out of any two *one-ers*. You can make an ortho-pair out of a *one-er* and a *two-er* by finding a *two-er* in which the grey level of the *one-er* is replaced by a white level in the *two-er*.

You can make an ortho-pair out of two *two-ers* by having two grey levels in one of the pair and making sure it is spotted at the same levels in the other of the pair.

You can make an ortho-pair out of a *two-er* and a *three-er* by making sure that if your *two-er* has two grey levels, its mate among the *three-ers* should also be grey at the same levels; but if your *two-er* has two spotted levels, its mate among the *three-ers* should be spotted at one of these levels and grey at the other of these levels.

You cannot make ortho-pairs out of two *three-ers*, nor out of a *one-er* and a *three-er*.

If you check the axes on your cube corresponding to an ortho-pair, you will see that the two axes are at right angles to each other. In other words they are orthogonal to each other. Hence the name ortho-pair.

Table 13 shows some ortho-pairs, which you should check, just in case I have made some mistakes.

Finally, here are some suggestions about how you can twiddle your cube so that you can generate the permitted sets of numbers by suitable twiddles.

Hold the cube by two opposite faces, placing your right index finger on the centre of one face and the left index finger on the centre of the opposite face Then use your two middle fingers to turn the cube round. You will get to see, one after the other, all the numbers between your hands that do not have the number that you are holding with your index fingers as a factor. Two of these lie on an edge, two of them at the centre of a face.

Table 13

	3	5	2	10	6	15	100	36	225	90	150	60	30
3		Orth	Orth	Orth			orth						
5	Orth		Orth		Orth			Orth					
2	Orth	Orth				orth			Orth				
10	Orth						orth				Orth	orth	
6		Orth						Orth		Orth		orth	
15			Orth						Orth	Orth	Orth		
100	Orth			Orth						Orth			Orth
36		Orth			Orth						Orth		Orth
225			Orth			orth						orth	Orth
90					Orth	orth	orth						
150				orth		orth		Orth					
60				orth	Orth				Orth				
30							orth	Orth	Orth				

(ii) Hold the cube at the mid-points of opposite edges, starting with these edges in a vertical position. It is best to use your thumbs and middle fingers, and do the twiddling with your index finger.

If you are holding edges with a square number and start twiddling slowly, you will see at the start the prime number in the centre of a face that is not a factor of your square number, and as you go on 'twiddling', the three other numbers which have the factors of your square number, but never squared.

If you are holding edges with a non-square number, you will still see the number that is not a factor of the number on the edges you are holding at the start, and as you twiddle you will see numbers with the same factors as the one on the edges you are holding, but one or the other factor or both will be squared.

(iii) Hold the cube at two opposite vertices with thumb, index finger and middle finger and twiddle. You will see three numbers turning up between your hands one after the other, each time taking in two of the three factors of your vertex number. If one of these factors occurs squared in your vertex number, then you will get a non-square number with this factor as one of its factors. If you consider the two unsquared factors in your vertex number, the product of these will occur squared on one of the edge numbers.

In other words, putting it plainly:

(i) twiddling about the 30 axis, you will see appearing 36, 100, 225 on the edges
(ii) twiddling about the 60 axis, you will see appearing 6, 10, 225 on the edges
(iii) twiddling about the 90 axis, you will see appearing 6, 15, 100 on the edges
(iv) twiddling about the 150 axis, you will see appearing 10, 15, 36 on the edges

The numbers that thus appear are the end points of axes belonging to one and the same plane, each of them at right angles to the axis about which you are twiddling. In the first nine cases these planes are planes of symmetry of the cube, in the last four cases they are end points of axes joining mid-points of opposite edges, thus cutting the cube in two equal but non-symmetrical halves in a hexagonal cut.

Chapter X
THE TENNIS PARTIES PROBLEM

1. The Tennis Parties Problem

The following thirteen people want to play tennis in foursomes. They do not mind with whom or against whom they play, but they want to be in as many teams of four so as they will have played in the same team with every other member of the party, but only once. These are the names of the players: Adam, Ann, Betsy, Bill, Bruno, Carol, Charlie, George, John, Karen, Laurie, Mary and Michael.

Michael seems to be the boss and makes some additional requirements. Ann, Betsy and Carol do not play well together in a team, so Michael says they should never be all three together in the same foursome. The same is true of Adam, Bill and Charlie.

But Michael insists that he wants to play in a team in which Ann and Adam are included, also in another team in which Betsy and Bill are included, and finally in a team in which Carol and Charlie are included.

Now you make up the teams, so that everyone teams up with each of the other players but only once. Don't forget to satisfy Michael's requirements as well!

When you have made up your teams (there should be thirteen teams), then work out who the following players are:

(i) The person who plays in a team with Ann and Betsy, as well as in a team with Adam and Bill.

(ii) The person who plays in a team with Carol and Betsy, as well as in a team with Charlie and Bill.

(iii) The person who plays in a team with Carol and Ann, as well as in a team with Adam and Charlie.

Do the three players you have found ever play in the same team? And if so, who is the fourth player in that team? Find an umpire to be responsible for each game in such a way that if any three games have a player in common, the three respective umpires should play in one and the same team. An umpire can or need not belong to the team that is playing.

If you turn to the next page, you can see a solution to the problem.

2. A Suggested Solution to the Tennis Parties Problem

Table 14 shows the thirteen teams that satisfy all the conditions of the problem. The umpires are placed under each team, in parentheses.

Table 14

Michael, Adam, Ann, Bruno	Michael, Bill, Betsy, George	Michael, Charlie, Carol, John
(Adam)	(Laurie)	(Carol)
Michael, Karen, Laurie, Mary	Adam, George, Carol, Laurie	Adam, Charlie, Betsy, Mary
(George)	(Michael)	(Bruno)
Adam, Bill, Karen, John	Ann, John, Betsy, Laurie	Ann, Bill, Carol, Mary
(Ann)	(Bill)	(John)
Charlie, Karen, George, Ann	Bruno, Betsy, Carol, Karen	Bruno, Bill, Charlie, Laurie
(Karen)	(Charlie)	(Betsy)
Bruno, George, John, Mary		
(Mary)		

The persons under (i), (ii) and (iii) are John, Bruno and Mary. They are in the last team listed, and the fourth player in their team is George. You will also find that the three umpires of any three teams with a common player also play in one and the same team.

When you have made yourself familiar with the above problems, then you can study Desargues's theorem, but only if you want to know the jargon. Here it is:

Consider a triangle ABC and another triangle A´B´C´, both in the same plane, but such that the lines AA´, BB´ and CC´ meet at the point M. The intersection of AB and A´B´, the intersection of BC and B´C´ and the intersection of AC and A´C´ all lie on the same straight line.

In the tennis problem the points are the players, and the lines are the teams. The one player common to two teams is the point at which the corresponding lines intersect. The team determined by any two chosen players corresponds to the line determined by the two corresponding points. Or you could call the teams the points and the players the lines, it would not make any difference. This is what mathematicians call duality.

Of course, my solution is only one of many. Sit down on a rainy afternoon and try to find other solutions!

3. The Construction of the Cross-Axis

By now Adam, Bill and Charles have made it up, and so have Anne, Betsy and Carol, and have now elected to play together. Michael still wants to be in one and the same team with Adam and Ann, and he also wants to be in the team in which Bill and Betsy play, as well as in the team in which Carol and Charles play. The teams should now be made up and you should see who the following players are:

(i) The player who is in the Adam–Betsy as well as in the Ann–Bill team
(ii) The player who is in the Bill–Carol as well as in the Betsy–Charles team
(iii) The player who is in the Adam–Carol as well as in the Ann–Charles team

Are these three players playing in one and the same team? Allot an umpire to each team as before, so that any three teams with a common player will have three umpires that are in the same team.

Table 15 shows the new teams, the umpires being shown as before. Since the teams are different, so are the umpires, although everyone gets a chance to be umpire as there are thirteen players and there are thirteen teams:

Table 15

George, Adam, Bill, Charlie	George, Ann, Betsy, Carol	George, Mary, Laurie, John
(Bill)	(Bruno)	(Mary)
George, Karen, Bruno, Michael	Adam, Betsy, Mary, Karen	Adam, Carol, Laurie, Bruno
(Ann)	(Laurie)	(Betsy)
Adam, Ann, John, Michael	Bill, Ann, Mary, Bruno	Bill, Betsy, Laurie, Michael
(Michael)	(George)	(Adam)
Bill, Karen, Carol, John	Charlie, Betsy, Bruno, John	Charlie, Carol, Mary, Michael
(Charlie)	(Carol)	(John)
Charlie, Ann, Laurie, Karen		
(Karen)		

When you have worked out this problem for yourself, or at least checked whether my solution above is correct, then you can begin to understand how to construct a cross-axis. Ignore this geometry if more fun and not more mathematics is your aim.

If the three points A, B and C are in line and if the points A´, B´ and C´ are also in line (but not in the same line), then let the two lines meet at a point G. If the lines AA´, BB´ and CC´ all intersect at the same point M, the intersections AB´ with BA´, BC´ with CB´, AC´ with CA´ are all in line. This line is called the cross-axis of the two ranges of points A, B, C and A´, B´, C´.

4. A Different Way of Formulating the Problem

Instead of thirteen players, you could have different elements in terms of which to formulate the problems. Let us construct thirteen elements out of three letters A, B and C with a slash separating the letters from each other. The order in which the letters are placed is immaterial, but I will try to keep to an alphabetical order, except in one case. Here are the elements of the game:

A, B, C, A/B, B/C, A/C, AB, BC, AC, ABC, A/BC, AB/C, B/AC

Let us decide on these four reasons for considering that four of the above elements belong to a foursome:

(i) One of the letters is missing.
(ii) Two of the three letters are always on the same side of the slash or neither letter is present.
(iii) Two of the three letters are always separated by a slash or neither letter is present.
(iv) There is one three-letter element and there are three two-letter elements in the set, but any two letters that are separated in the three-letter element are not separated as they occur in the two-letter element; also any two letters that are not separated in the three-letter element are separated as they occur in the two-letter element.

Here are the foursomes that can be made according to the above rules:

B, C, BC, B/C
A, C, AC, A/C
A, B, AB, A/B
BC, ABC, A/BC, A
AC, ABC, B/AC, B
AB, ABC, AB/C, C
B/C, AB/C, B/AC, A
A/C, A/BC, AB/C, B
A/B, A/BC, B/AC, C
AB/C, A/B, AC, BC
A/BC, AB, AC, B/C
B/AC, AB, BC, A/C
ABC, A/B, B/C, A/C

You might like to check the following interesting facts about these thirteen foursomes:

(i) Any two foursomes have one and only one element in common.
(ii) Any two given elements belong to just one foursome.
(iii) Every foursome has four elements.
(iv) Every element belongs to four different foursomes.

5. Duality

We could now try to associate an element to each foursome but in such a way that to any three foursomes having a common element should correspond respectively three elements of one and the same foursome. This is much easier to do in this embodiment of the problem than in the case of choosing umpires for our tennis matches.

- (i) To each foursome in which a letter is missing, let us associate the missing letter.
- (ii) To each foursome characterised by a pair of 'undivided' letters, let us associate the same pair of letters, but with a slash between them.
- (iii) To each foursome characterised by a pair of 'divided' letters, let us associate the same pair of letters, but without a slash between them.
- (iv) To each foursome with one three-letter element and three two-letter elements, let us associate the three-letter element which is a member of that same foursome.

You should check that this solves the umpire problem.

You must have noticed that I have always picked an associated element to every foursome which is in a sense opposite to the elements of the foursome. If a letter is missing, then this same missing letter has been picked. For undivided foursomes a divided element has been chosen. For divided foursomes an undivided element has been chosen. Even in the last cases, the two-letter elements are in a sense opposite to how the two corresponding letters are placed in the three-letter element.

6. Links with the Axes of the Cube

I think it is time I gave the game away. If you have studied mathematics, what follows will be no surprise. So to explain where all these problems come from, I should like you to get hold of a cube, or you could make one out of cardboard. Have a black and a red pencil ready. Draw the letter A with a black pencil on one face, then draw the letter A with the red pencil on the opposite face. Draw a black B on another face and a red B on the face opposite to this one. Then draw a black C and a red C on the remaining two opposite faces.

The line joining the centres of the A faces is the A axis. The line joining the centres of the B faces is the B axis and the line joining the centres of the C faces is the C axis.

There are twelve edges on the cube, therefore six pairs of opposite edges. The mid-point of each edge can be joined to the mid-point of the opposite edge. These are six more axes. We can use the letters on the adjacent faces as the names of these axes. There will be a slash between the letters if the letters on the adjacent faces are of different colour, and there will be no slash if the letters on these adjacent faces are of the same colour.

The four axes joining the opposite vertices will use all three letters. There will be one such axis where all the concurrent faces have the same colour. This is the ABC axis. The other axes will have concurrent faces around them, two of them having the same colour and the third one having a different colour. The letter which has the different colour then will be separated by a slash from the other two letters.

So each of the thirteen axes of rotation of the cube will have acquired a name. And these names are the thirteen elements of our second game.

You might like to check that in the case of nine foursomes out of the thirteen foursomes, the four axes of the same foursome all lie in the same plane. What is more, the umpire element is an axis at right angles to this plane. In the case of the foursomes in which there is one three-letter element and there are three two-letter elements, the axes corresponding to the two-letter elements all lie in the same plane, which is perpendicular to the axis corresponding to the three-letter element. These are the only four foursomes or teams of tennis players in which the umpire belongs to the foursome or tennis team.

The plane determined by the axes corresponding to the two-letter elements intersects the surface of the cube in the shape of a regular hexagon. The other planes intersect the cube in a rectangle (which is a square in the case of the foursome with one letter missing).

Questions:

(i) In the two problems, namely Desargues's theorem and the cross-axis, if the three points in question at the end are in line, does it follow that the joins of the corresponding points all meet at a point?

(ii) Can you prove Desargues's theorem by using only incidence properties (namely about joins and intersections)?

(iii) What happens if you cheat and introduce a third dimension, so that two planes meet on a line, three planes that do not meet on a line meet at a point, three points determine a plane, unless they are in line, etc.? Have the triangle ABC in one plane, the triangle A´B´C´ in another plane, and the point in which AA´, BB´ and CC´ meet at M will not be in either plane. Then AB and A´B´ must meet in the line of intersection of the two planes since AA´ and BB´ are coplanar because they meet at M! The same applies to AC and A´C´ as well as to BC and B´C´. So Desargues's theorem is true in three dimensions. Can you use this fact to prove Desargues's theorem for two triangles in the same plane?

If you find the umpires of four tennis teams that have a common player, and locate the single team in which these umpires play, can you tell right away who the umpire of that single team is?

Chapter XI

DISC SEQUENCES.

1. Introducing the Discs

In what follows we shall be using discs, each of which is divided into three equal sectors. On each sector there are some pictures of coconut trees and starfish. Some are thin, some are dotted and some are thick. Of each kind in each sector there could be either nothing or a coconut tree or a starfish or one of each.

This gives us four possibilities. Since there are three kinds, it means we can make pictures for 64 sectors. If we do not count the picture with absolutely nothing on it, that leaves us with 63 pictures. Since there are three sectors on each disc, it means we can fill 21 discs, no two sectors being exactly alike.

The starfish and the coconut trees are arranged on each disc in such a way that for each kind the cyclic order:

starfish → coconut tree → both → starfish

is observed as we go round the disc in the clockwise sense. Figure 62 shows a sample disc:

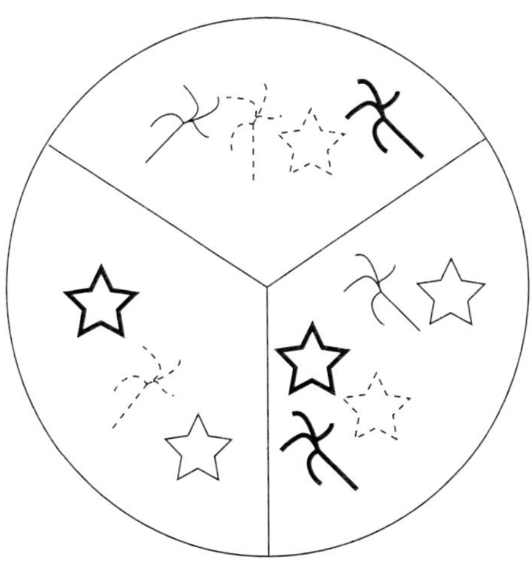

Figure 59

To describe what there is in any particular sector, we shall use the symbol 0 if a kind is missing, the symbol 1 if there is just one starfish, the symbol 2 if there is just one coconut tree and the symbol 3 if there is a starfish and a coconut tree of that kind. We shall record what is in each sector in this order: thin, dotted, thick. Our sample disc will then be described like this (see Figure 60).

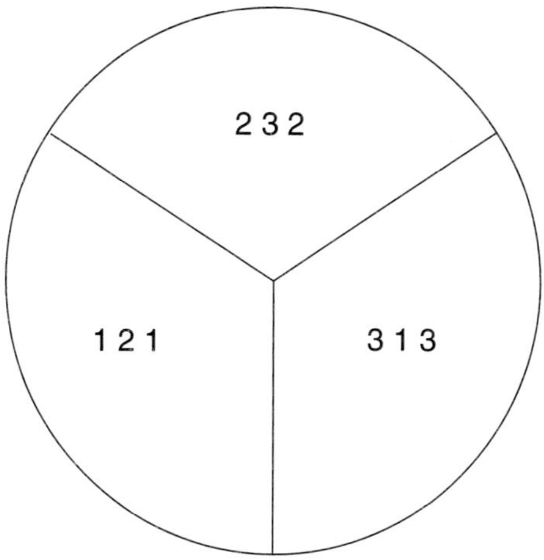

Figure 60

2. Addition and Multiplication of Sectors of the Discs

As we have seen, each sector is put together out of four different elements: nothing, starfish, coconut tree and both, in each of the three kinds thin, dotted and thick. If we want to do anything with these elements, we have to invent some ways of relating them to each other. One thing we can do is to define how to add our elements to each other. The easiest suggestion is the following:

Adding any two of the four elements, we merely put them together, but if by doing so we find two of anything, we cross them out. For example, a starfish added to a starfish and a coconut tree would be a coconut tree. Putting these elements together, there would be two starfish which we would have to cross out.

It is easy to verify that our adding table will be as shown in Table 16.

Table 16

Add	0	1	2	3
0	0	1	2	3
1	1	0	3	2
2	2	3	0	1
3	3	2	1	0

Now we know how to add. Naturally we shall be adding thin ones to thin ones, dotted ones to dotted ones and thick ones to thick ones. For example:

3 1 1 + 2 1 3 = 1 0 2

The thin ones have lost a coconut tree, the dotted ones have been reduced to nothing, as the two starfish have had to be cancelled, and the thick ones have also lost two starfish, there being just a coconut tree left. It must be remembered that these numbers have nothing to do with the usual numbers 0, 1, 2 and 3. These symbols are being used in the way described just to have a way to describe what there is in a picture. We could equally well have used any other symbols such as letters or even the pictures themselves.

If you do not go for starfish and coconut trees, you could draw boys and girls, or cups and saucers, or knives and forks or anything else that you fancy. All you have to be sure of, so far, is that you know which of your symbols behaves like 0. It is best to use the nothing there!

Now let us have a shot at inventing a multiplication:

(i) Multiplying any of the elements by 0 should result in 0. Why rock the boat and have it too fancy?

(ii) Multiplying by 1 could simply leave any element unaltered. After all, this is what happens when you normally multiply by 1. Again, why make it too hard?

(iii) Multiplying by 2 should turn 1 into 2 (so as to keep the usual property of neutrality of 1), so why not make 2 turn 2 into 3 and so 3 into 1? So multiplying by 2 will make us take one step round the cycle 1 → 2 → 3 → 1. Naturally 0 × 2 should be 0.

(iv) Multiplying by 3 should turn 1 into 3 (for reasons of 1's neutrality), so why not make it turn 3 into 2 and 2 into 1? So multiplying by 3 will make us take one step round the cycle 1 → 3 → 2 → 1. Naturally 0 × 3 should be 0.

The multiplication table will then look like Table 17.

Table 17

Multiply	0	1	2	3
0	0	0	0	0
1	0	1	2	3
2	0	2	3	1
3	0	3	1	2

One good way of remembering the addition table is by remembering the following:

(i) Combining any element with 0 gives us that same element.

(ii) Combining any two of the elements 1, 2, 3 always gives us the third one.

(iii) Combining an element with itself gives us 0.

To remember the multiplication table we should try to remember the following:

(i) Combining any element with 0 gives 0.
(ii) Combining any element with 1 gives that same element.
(iii) Out of the two elements 2 and 3, combining one of them with itself gives the other, combining them with each other gives 1.

We can multiply a whole sector by an element by simply multiplying each kind of the sector by our multiplier element. For example:

$$(2\ 3\ 2) \times 2 = (3\ 1\ 3)$$

$$(2\ 3\ 2) \times 3 = (1\ 2\ 1)$$

It will be noticed that multiplying by 2 takes us to the next sector on the disc in the clockwise sense, whereas multiplying by 3 takes us to the next sector on the disc in the counterclockwise sense.

3. Defining Properties that Some Discs Have and Others Do Not Have

Let us look at our discs now and see what interesting properties they might have. We might classify our discs by noting whether they have or do not have such properties (see Figure 61).

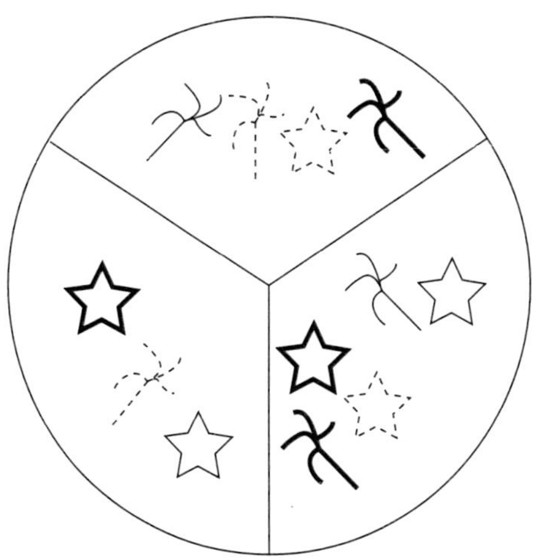

Figure 61

There are some very obvious properties we can think of right away. For example, (ntn) or no thin, (nd) or no dotted, (ntk) or no thick. Sometimes in every sector, whatever there is in one kind, there are the same elements in another kind. These could be the properties (tnd) or thin = dotted, (dtk) or dotted = thick, (tktn) or thick = thin.

Sometimes we might find that what we have of one kind in one sector we have in the next sector counterclockwise but in another kind. This gives us six more properties, which we could symbolise as follows:

(tn → d), (d → tk), (tk → tn), (d → tn), (tk → d), (tn → tk)

For example, it is not hard to think of a disc which has the following three properties:

(tn → tk), (d → tk), (tnd)

All we have to do, for example, is to put a thin star and a dotted star in one sector, then by the rules put a dotted star and tree and a thin star and tree in the next sector counterclockwise and a dotted tree and a thin tree in the remaining sector. But to obey the first two properties, we shall have to put a thick tree in our first sector, followed by a thick tree and star in the next one clockwise, followed by a thick star again clockwise.

How many more properties can we think of? One way to make some decisions might be to find out how many discs do and how many discs do not have any one of the properties so far thought out. It might surprise you to know that the answer so far is always five discs. So it might be a good idea to try to think of other properties that are possessed by exactly five discs.

We could think of a property such as (d → tn) as one in which adding the dotted element in one sector to the thin element in the next sector in the counterclockwise sense gives us zero, elements of the same kind having to cancel out. So we might think of different ways of adding up elements of different kinds in different sectors of the same disc, and seeing whether we get them all cancelling out.

Let us use the slash as a symbol to separate what we need to look at in one sector from what we need to look at in another sector. Whatever comes after this is to be looked at in the next sector going round in the counterclockwise sense. Doing this we can propose the following further properties:

(tnd/tk), (tk/tnd), (dtk/tn), (tn/dtk), (tktn/d), (d/tktn), (tk/tn/d), (tn/ttk/d), (gbr)

the last property meaning that all the three kinds should be counted which are in the same sector.

4. Working with Properties and Sequences

It can be verified that each of these properties yields exactly 5 discs. It is also true that each disc will have exactly 5 of the above 21 properties. Any two discs will have just one property in common. Any two properties will be possessed by just one of the twenty-one discs. If you do not believe all this, make the discs yourself and check out the above statements!

Below are descriptions of five discs with the property (dtk/tn). I shall denote the presence of a tree and a starfish of the same kind as double, so there will be trees, stars and doubles. I shall use the order thin, dotted and thick, and a comma will mean 'go to the next sector counterclockwise'. I shall write zero for 'nothing of that kind'.

(star tree star, double star double, tree double tree)	=	(121, 313, 232)
(tree star zero, star double zero, double tree zero)	=	(210, 130, 320)
(zero tree tree, zero star star, zero double double)	=	(022, 011, 033)
(double zero tree, tree zero star, star zero double)	=	(302, 201, 103)
(double double star, tree tree double, star star tree)	=	(331, 223, 112)

Take the last two elements from any sector and add the first element in the sector you reach by going counterclockwise. You will always get either two stars or two trees or two of both.

You can try something else: take any two of the fifteen sectors in the above set of sectors and add them. The sum will always be one of the fifteen sectors. So we have a closed system. If you put together any other five discs which all have one property in common, you will find another closed system; any adding will never get you out of the set of fifteen sectors that are on the five discs having your chosen common property.

There is yet another interesting feature of our five discs and it consists in the order chosen to place them. There are rules we can find which will give us one sector of any of the discs from a certain sector of the disc just before it, excepting of course the first disc. One rule could, for example, be like this, using the multipliers:

(i) 0, meaning that there is no element of that kind
(ii) 1, meaning that we take the element of that kind as it is
(iii) 2, meaning that for that kind we replace star by tree, tree by double and double by star (1 to 2 to 3 to 1)
(iv) 3, meaning that for that kind we replace star by double, double by tree and tree by star (1 to 3 to 2 to 1)

thin in next sector = 1 thin + 2 dotted + 0 thick
dotted in next sector = 0 thin + 2 dotted + 2 thick
thick in next sector = 1 thin + 0 dotted + 1 thick

For example, the sector after (1 thin star, 1 dotted tree, 1 thick star) or (121) would be calculated like this:

thin in next sector = $(1 \times 1) + (2 \times 2) + (0 \times 1) = 1 + 3 + 0 = 2$
dotted in next sector = $(0 \times 1) + (2 \times 2) + (2 \times 1) = 0 + 3 + 2 = 1$
thick in next sector = $(1 \times 1) + (0 \times 2) + (1 \times 1) = 1 + 0 + 1 = 0$

Which means a thin tree and a dotted star and no thick elements.

Our starting sector is the one referred to as the first sector on our first disc. Underneath this, in the same column, we shall find the sector containing a thin tree and a dotted star and no thick elements.

If we apply the rule to the fifth disc, we come back to the first disc, but of course to another sector on it. In fact, we obtain the third sector referred to on our first disc. Then we go down this column if we go on applying our rule, eventually doing all three columns and after fifteen applications of the rule, we shall be back at our very first sector.

Our rule is a rule of order 15. We have to use it fifteen times in succession before returning to our starting point.

5. Arranging the Discs in a Four by Four Pattern

One interesting problem is to arrange 16 of the 21 discs in 4 rows of 4 discs. This should be done in such a way that the discs in one and the same row all have a common property, as well as so that the discs in one and the same column all have a common property. This is quite a challenging problem. If you want to try it without any hints, do not read any further.

Hint (1) Which five discs do you think you would need to leave out of your set so as to make the problem possible to solve?

Hint (2) Try taking out five discs all having a common property.

Hint (3) Choose two of the discs that you have taken out. Each of these will have four other properties besides the one that is common to all five. How do you think these properties will help you solve the problem?

Suppose we take out the five discs with the property (tndtk), in other words those discs on which each sector has either two or no starfish and either two or no coconut trees. Let us pick the discs, one of which has the sector (213) while the other has the sector (231). These discs have the following properties in addition to the property (tndtk) common to all the five discs:

(tn/d/tk), (tn → tk), (tk → d), (d → tn) for the first one
(tn/tk/d), (tk → tn), (tn → d), (d → tk) for the second one

We can now arrange our sixteen discs so that our rows will have the properties of our first disc and our columns will have the properties of our second disc (see Figures 62, 63 and 64).

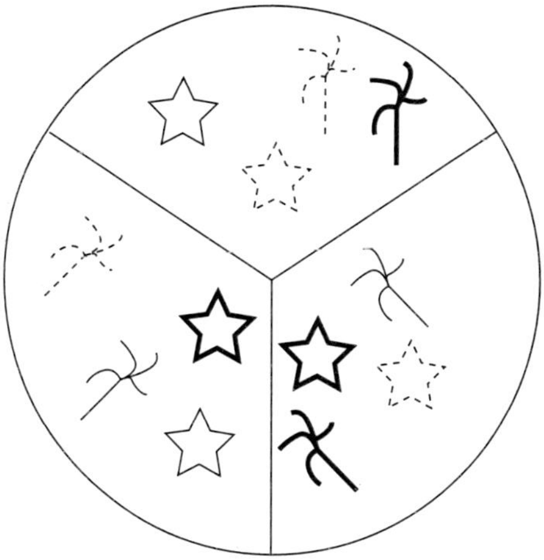

first disc

Figure 62

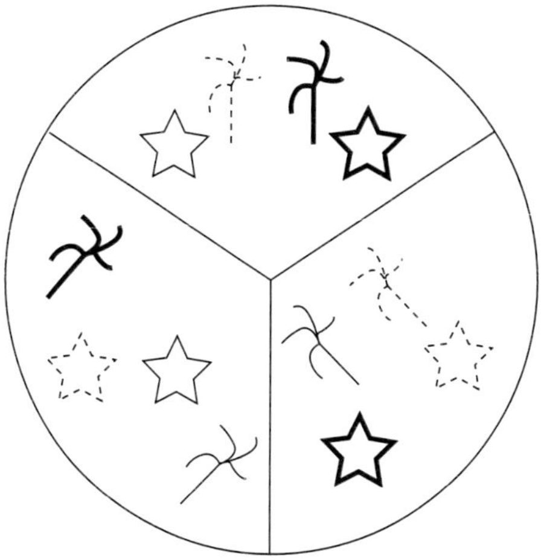

second disc

Figure 63

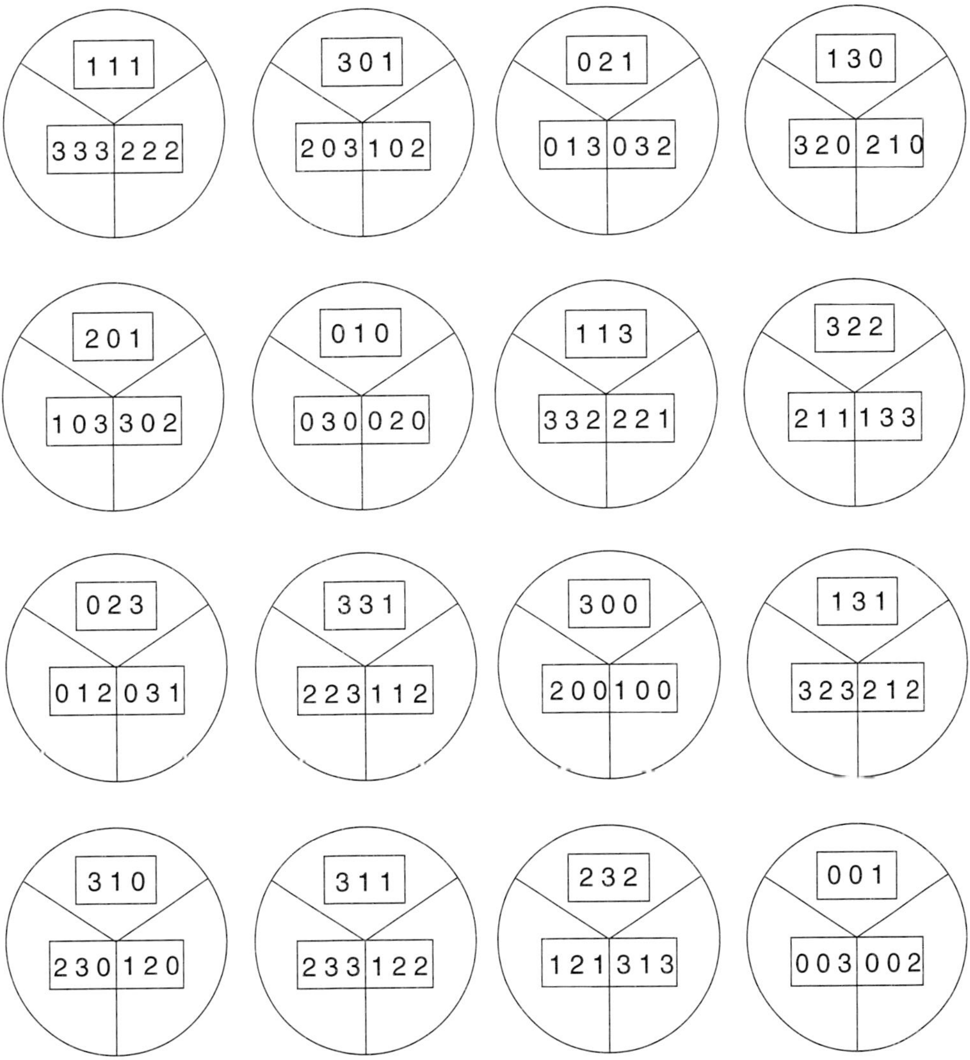
111
333 222
301
203 102
021
013 032
130
320 210
201
103 302
010
030 020
113
332 221
322
211 133
023
012 031
331
223 112
300
200 100
131
323 212
310
230 120
311
233 122
232
121 313
001
003 002

Figure 64

Remember to read the numbers from left to right as meaning first the thin element, then the dotted one and then the thick one. The number 1 means a star, the number 2 a coconut tree and the number 3 means that there is one of each. Remember not to confuse walking round the disc clockwise with turning the disc clockwise. One of these will always produce the opposite sense of turn of the other.

6. Where Are the Discs which Have the Properties of the Remaining Three Discs out of the Five we have Picked?

We have used up two of the discs with the property (tndtk): one for the rows of our square and one for its columns. You might begin to wonder where the discs are that use up the properties of the remaining three discs? Are they just randomly scattered all over the square or is there some discernible pattern? If there is no recognisable pattern, is there a way of reorganising the discs so as to create such patterns? If so, what kind of patterns are possible? Will they remain invariant under rotation or under reflection?

These are questions you could well ponder one day when you are forced to stay at home because of a snowstorm. But as you will see from Figure 65 there are some patterns that I have been able to get for the remaining discs.

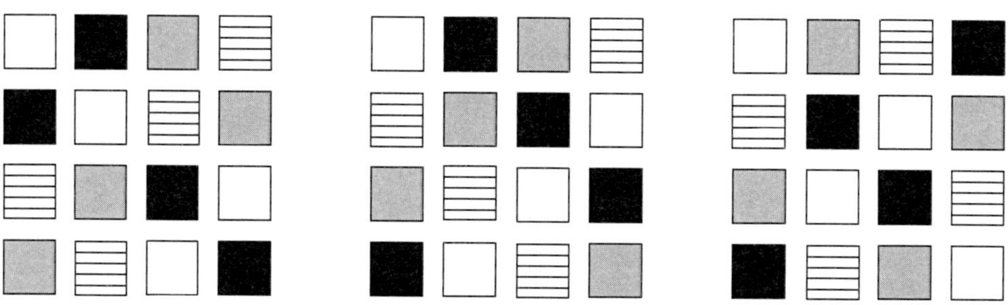

Figure 65

Each pattern shows four discs with a common property. You can see for yourself which of these will always show four discs with a common property even if turned through a right angle or flipped. If you can come up with some better patterns, I should love to hear from you!

7. How to Generate Sequences of Sectors through Rules

You can make a lot of different sequences, using different rules for generating them. Here is one, the rule being given algebraically:

next thin = 1 × thin + 1 × dotted + 2 × thick
next dotted = 1 × thin + 2 × dotted + 1 × thick
next thick = 1 × thin + 1 × dotted + 0 × thick

You will find that this will take up seven discs. The sequence will go through all the pictures on all the seven discs. If you start the sequence on a picture which is not on any of these seven discs, you will get another sequence on seven other discs. In this way you can make up three sequences, each containing 21 pictures. This means you have used up all the pictures!

Table 18 shows the kind of thing you are likely to get.

Table 18

I	II	III
1 3 0	3 3 2	0 3 2
2 0 2	3 0 0	0 3 3
1 0 2	3 3 3	2 2 3
2 3 1	1 1 0	1 2 0
3 2 1	0 3 0	3 2 3
3 1 1	3 1 2	0 3 1
0 0 2	3 2 2	1 0 3
3 2 0	2 2 1	0 2 1
1 0 1	2 0 0	0 2 2
3 0 1	2 2 2	1 1 2
1 2 3	3 3 0	3 1 0
2 1 3	0 2 0	2 1 2
2 3 3	2 3 2	0 2 3
0 0 1	2 1 1	3 0 2
2 1 0	1 1 3	0 1 3
3 0 3	1 0 0	0 1 1
2 0 3	1 1 1	3 3 1
3 1 2	2 2 0	2 3 0
1 3 2	0 1 0	1 3 1
1 2 2	1 2 1	0 1 2
0 0 3	1 3 3	2 0 1

I am tempted to show you some interesting regularities about the above. For example, if out of any run of ten pictures you add the first, the fourth and the tenth pictures, adding always the same kind together, you will get 0 for each kind. We could call three such pictures a triad.

You can find other triads, using more than one of the three sequences. For example, the nth picture and the (n + 1)th picture in the first column will make a triad with the nth picture in the second column. This will also be true of the second and the third columns. Also the nth and the (n + 2)nd pictures in the first column will make a triad with the nth picture in the third column.

You might want to see how many of the 21 properties are used by any one sequence. The answer is they are all used. Some are used by only one disc, others by three discs. No property is possessed by any other number of discs but by either one or three.

You might have noticed that out of any run of four discs the first, the third and fourth discs share a common property. This could be used as an amusing exercise for the next snowstorm. The problem would be to pick seven discs and place them in a circular pattern, a picture on each disc always pointing towards the centre of this circle, so that out of any run of four discs, the first, the third and fourth discs should share a common property. It will be necessary for the pictures on the first, the third and the fourth discs that point towards the centre of the circle of discs to form a triad. Players might need to change some of the discs before they succeed in this feat. If the cycle has been successfully completed, there will always be a rule that will determine the next picture round the cycle. One such rule could be, for example:

next thin = $2 \times$ thin + $0 \times$ dotted + $1 \times$ thick
next dotted = $1 \times$ thin + $1 \times$ dotted + $2 \times$ thick
next thick = $0 \times$ thin + $2 \times$ dotted + $3 \times$ thick

If according to the above rule the eighth picture is the same as the first picture, then the rule is of order 7. If on the other hand, the eighth picture is another picture on the first disc, not the one pointing to the centre, then the rule will take you round all the pictures of all the seven discs, the rule being then of order 21.

How about looking at the following rule?

next thin = the present thick
next dotted = (the present thin) + (the present thick) $\times$ 2
next thick = the present dotted

The above rule is easier to explain in terms of how the elements are placed on the discs. We can say:

(i) The thin element in the next sector is the same as the thick element in the sector we are looking at.
(ii) The dotted element in the next sector is obtained by putting together the thin element of the sector we are looking at, added to the thick element on the left of this sector.
(iii) The thick element of the next sector is the same as the dotted element in the sector we are looking at.

You will find that if you make the sequence generated by the above rule, out of any run of four discs in the sequence the first, second and fourth discs share a common property.

I will leave you with the $64,000 question. Is there a rule that takes in all 63 sectors in one and the same sequence? If so, what is it?

You might try the following rule for generating a sequence:

From one sector	to the next sector
↓	↓
dotted element	thin element
thick element	dotted element
sum of all elements but changed along the cycle star to tree to double to star	thick element

Chapter XII

PRISM GAMES WITH TWENTY-ONE PRISMS

Figure 66 shows a plan to show you how to make the prisms. S means a star and T means a tree and D means a tree and a star, namely a double, and O means that there is nothing there.

<table>
<tr><td>O O S
O O T
O O D</td><td>O S S
O T T
O D D</td><td>S O S
T O T
D O D</td><td>S S O
T T O
D D O</td><td>S S T
T T D
D D S</td><td>S S D
T T S
D D T</td><td>S S S
T T T
D D D</td></tr>
<tr><td>O S O
O T O
O D O</td><td>O S T
O T D
O D S</td><td>S O T
T O D
D O S</td><td>S T O
T D O
D S O</td><td>S T S
T D T
D S D</td><td>S D S
T S T
D T D</td><td>S T D
T D S
D S T</td></tr>
<tr><td>S O O
T O O
D O O</td><td>O T S
O D T
O S D</td><td>T O S
D O T
S O D</td><td>T S S
D T T
S D D</td><td>T S O
D T O
S D O</td><td>D S S
S T T
T D D</td><td>S D T
T S D
D T S</td></tr>
</table>

Figure 66

(i) Copy each of the above boxes onto a rectangular piece of cardboard divided into four equal rows, drawing the first picture in the top row, the middle picture in the second row and the third picture in the third row, putting nothing in the fourth row.

(ii) Use green for the left hand part of each picture, yellow for the middle part and red for the right-hand part.

(iii) Any part of a picture where an O is shown should be left without drawing anything there.

(iv) Make three folds between your four rows, and make the prism by sliding the empty fourth row under the first row. This will make a triangular prism with your pictures on the three faces.

(v) You could stick the first and fourth parts of your cardboard rectangle together with glue.

When you have made your 21 prisms with a picture of trees and/or stars on each of the three faces of every prism, just check on the following:

(i) If you roll one of your prisms towards you, you will notice that a star is replaced by a tree of the same colour, a tree is replaced by a tree and a star of the same colours, and a tree and a star of the same colour are replaced by a star of that colour.

(ii) If you roll your prism away from you, you will notice that a star is replaced by a tree and a star, a tree and a star (of the same colour) are replaced by a star and a star by a tree, always of the same colour.

(iii) Three prisms will have faces with only one colour, nine prisms will have two colours and another nine prisms will involve all three colours.

For the sake of brevity, I shall call a star and a tree of the same colour a double. I shall also refer to rolling towards you as multiplying by tree or by T for short. Rolling away from you will be also called multiplying by double or by D for short. Multiplying by star will mean leaving the prism as it is: it will be a multiplication by 1.

I shall sometimes use O to indicate that there is no element of a certain colour in a picture on a prism. This can be thought of as an upper case O if letters are used or as 0 zero if numerals are used. I shall sometimes use 1 for a star, 2 for a tree and 3 for a double. I shall always refer to the green first, then the yellow and then the red, as this is the order in which they appear from left to right on each picture. So, 302 or DOT would mean that there is a green double, that there is no yellow and that there is a red tree on the picture referred to.

I am indebted to Roy Davies of Leicester University for first suggesting a sequence of length 63 of which I have made use in what follows, and to George Rousseau, who used his computer skills and mathematical know-how to unearth all the other possible sequences of length 63 as well as some others, thus giving future players of the prisms games great opportunities for amazing their friends!

1. The Prisms Written Down in a Certain Order

You can use the sequence as shown in Table 19 for identifying any of the 63 pictures. You need to say whether your picture is in week 1 or in week 2 or in week 3, then say which day of the week has your picture, and finally you will have to say whether it is in a morning row or in a noon row or in an evening row.

Table 19

Sunday	Monday	Tuesday	Wednesday	Thursday	Friday	Saturday		
1TD	TDO	DOT	OTT	TTO	TOO	OOD	Morning	
TD1	D1O	1OD	ODD	DDO	DOO	OO1	Noon	Week 1
D1T	1TO	TO1	O11	11O	1OO	OOT	Evening	
OD1	D1D	1DT	DTO	TOT	OTO	TOD	Morning	
O1T	1T1	T1D	1DO	DOD	ODO	DO1	Noon	Week 2
OTD	TDT	DT1	T1O	1O1	O1O	1OT	Evening	
ODT	DTT	TT1	T1T	1TT	TTT	TTD	Morning	
O1D	1DD	DDT	DTD	TDD	DDD	DD1	Noon	Week 3
OT1	T11	11D	1D1	D11	111	11T	Evening	

2. Adding Pictures

You already know how to multiply a picture by 1 or by T or by D. Multiplying by 1 leaves the picture unaltered. We could say that multiplying by zero will get rid of your picture altogether. Multiplying by T takes you from morning to noon, or from noon to evening, or from evening to morning (of the same day).

Multiplying by D takes you from morning to evening, from evening to noon and from noon to morning (of the same day).

Adding two pictures is even easier. You put the parts of the same colour together, and if anything comes up twice in the sum, you leave it out. That means, for example, that a double and a tree (usually of the same colour) would be a star, since the two trees in the sum would have to be left out.. So we have the sums:

$$1 + T = D, 1 + D = T, T + D = 1$$

Of course:

$$1 + 1 = O, T + T = O \text{ and } D + D = O$$

You will find that any two pictures on the same prism will add up to the third picture on that same prism. Adding any two pictures on two different prisms will always get you a picture on some third prism.

Try adding the morning pictures of three consecutive days. You will get the evening picture of the fourth day. If you add two on consecutive days both at the same time, you will find the sum a week after the second day at the same time. Try to find other such features of the table.

3. An Adding Game

Choose any two prisms, and on each prism choose a picture. Add these pictures and find the third prism with that picture on. Let us say you have captured that third prism. Try to capture some more prisms by adding pictures on prisms already captured. Eventually you will find that adding does not capture any more prisms, since the picture which is your sum has already been captured.

This should happen when you have a total of five prisms. This is your first family of prisms. Give this family a name by looking to see in what way your pictures are like each other and unlike the others that have not been captured.

For example if neither of your starting prisms has any green on it, there is no way you can capture any prisms with green on. So you will have made the 'non-green' family. Or if both your starting prisms were such that on each picture the green part is just the same as the red part, then you will never capture any prisms on which the green part is different from the red part. In this case you will have made the 'green equals red' family.

Try to make up as many families as you can and try to give them suitable names. A name must refer to what is true for members of the family but false for any non-members! You are warned that any two families will always have one prism in common!

Here is an example of capturing. Suppose we choose the following prisms as shown in Table 20:

Table 20

O11	1TD	We have O11 + 1TD	1DT
		= 1DT so	
OTT and	TD1		T1D
ODD	D1T		DT1

has been captured! But O11 + TD1 = TTO so

TTO

DDO

1 1 O

has also been acquired. But since also O11 + D1T = DOD, we have acquired the prism

DOD

1O1

TOT

so as shown in Table 21 our family consists of:

Table 21

O11	1TD	1DT	TTO	DOD
OTT	TD1	T1D	DDO	1O1
ODD	D1T	DT1	11O	TOT

Adding any two of these fifteen we shall get one of the fifteen. Adding one to itself we get OOO which must belong to all families (the picture with nothing on it)! If we add everything that is in the same picture, we always get zero! So this is the zero family!

4. How Many Families Are There and What Shall We Call Them?

I might as well let you in on the fact that there are exactly 21 families that it is possible to make by starting with two prisms.

You will find the following names quite suitable. The word 'behind' refers to the picture facing away from you, which seen from where you are is behind the prism. 'Underneath' means the picture in contact with the table that you cannot see without lifting the prism. Here are the names:

(1) no green
(2) no yellow
(3) no red
(4) yellow = red
(5) red = green
(6) green = yellow
(7) yellow = red behind
(8) red = yellow behind
(9) red = green behind
(10) green = red behind
(11) green = yellow behind
(12) yellow = green behind
(13) red + yellow + green behind = zero
(14) red + yellow + green underneath = zero
(15) yellow + green + red behind = zero
(16) yellow + green + red underneath = zero
(17) green + red + yellow behind = zero
(18) green + red + yellow underneath = zero
(19) green + yellow behind + red under = zero
(20) green + yellow under + red behind = zero
(21) green + yellow + red = zero

If you choose one of these names and pick out the five prisms that have the property suggested by the name, you will always find that within the same family, the sum of any two pictures will always be on one of the prisms of that family. Or if you start with any two prisms and make a family of five prisms, one of the above names will always fit the family that you have made.

Any prism will belong to exactly five families, just as any family will consist of exactly five prisms. Any two prisms will belong to exactly one family and any two families will have exactly one prism in common.

5. Guess the Prism and Guess the Name Games

You could write the names on cards, thus making a deck of 21 cards. You could then play the following game:

The prisms and the cards are all spread out randomly on the table. One player picks two cards. The other players must try to find the single prism that belongs to both families defined by the chosen names. The first player to do so grabs the prism and so wins a point. Players then take it in turn to choose two cards, while the other players must concentrate on finding the correct prism that goes with the chosen cards.

You can also play the game by the first player picking up two prisms. The other players then try to find the name of the family of which both prisms are members.

Another way of playing it would be by the first player picking up three cards. In this case the answer from the other players would be either the prism that has all three names, if indeed there is such a prism, or, if there is no such prism, no such prism. If a player claimed the latter, he or she would have to show why no such prism could possibly exist!

Another way of playing this would be by the first player picking up three prisms. In this case the answer from the other players would have to be either family, accompanied by the name of the family to which all three prisms belong, or no such family, if indeed there is no family containing all three of the chosen prisms. A player claiming the latter would have to prove the point!

Here is an example. Suppose the first player draws the prisms:

OO1	O1D	1D1
OOT	OT1	T1T
OOD	ODT	DTD

The answer would be no such family, since the first two prisms belong to the 'no-green' family, while the third one clearly does not. So the third one is not a member of the family determined by the first two.

If someone chose the cards: green = red, green = yellow behind and yellow = green underneath, the answer would be the third of the above prisms.

6. Make a 4 × 4 Pattern with Sixteen Prisms

Choose 16 out of your 21 prisms and arrange them in four rows of four in such a way that the prisms in each row are members of one and the same family and also that the prisms in each column are members of one and the same family. You will find it easier to solve this problem if you make sure that the 5 prisms that you do not use all belong to one family!

In fact, you can use the unused prisms for solving the problem.

Choose one of the unused prisms and find its names besides the name it has in virtue of belonging to the family of unused prisms. There will be four such names. Call them A, B, C and D. Arrange your prisms in such a way that the prisms in row 1 all have the name A, the prisms in row 2 all have the name B, those in row 3 the name C and those in row 4 the name D. Now take another of your unused prisms and use its names for arranging the columns, but without disturbing the rows.

Then you might find it amusing to find the prisms that have the names of all the remaining three unused prisms! Do they make a pleasant pattern? If not, by changing rows and/or columns, can you improve the look of the arrangement? If you find this problem too vague, just go on to the next section. Anyway, as shown in Table 22, there is one possible solution for making the 4 × 4 pattern. Only one face of the prism is written down, as the other two faces can be deduced easily by tree-ing and by doubling.

Table 22

TTO	T1D	D1T	1O1
111	T11	D11	1OO
11T	T1O	D1D	1OD
11D	T1T	D1O	1OT

The unused prisms are the non-green ones.

You might like to check my work to see if I have made any mistakes. I believe, though, that each row can be completed by means of the prism with O11 on it and each column can be completed by means of the prism with O1O on it. There are three more non-green prisms. Find the names of each of these, and try to see how these names are distributed in the above diagram.

7. Make Up Rules for Sequences of Pictures

In what we have done up till now it did not seem to matter in what order we placed our prisms. Let me start you off with some ways of making sequences, in which the pictures must be placed one after the other in a certain order determined by some rule.

Here is one suggestion. Pick the picture with a green star and a yellow tree and no red on it for a starter. Then use the following rules for generating the next picture:

The green on the next picture will be whatever is yellow on the present one. The yellow on the next picture will be whatever is red on the present one. To get the red on the next picture add the green and the yellow on the present picture, and then add whatever is red behind on the present prism.

Your first picture is 1TO. So there must be a green tree and no yellow on the next picture. To get the red part we must find:

$$1 + T + (T \times O) = 1 + T = D$$

Since there is no red behind and star plus tree equals double, the next picture is TOD.

If you now continue, using the same rule but applying it to TOD, you will get ODD (how odd!), and after ODD you will get DDT (I thought we had got rid of DDT by now!) and so you will get this sequence:

1TO TOD ODD DDT DTD TDO DO1 O11 11D 1D1 D1O 1OT OTT TT1 T1T

after which you get back to 1TO and the whole thing starts again. Put your prisms out in this order.

You might have noticed some strange things about the sequence. For example, if you add any three consecutive pictures, then from the third picture you count on to the eighth picture, you will miraculously find your sum. Another feature of the sequence is that if you add the first and the third out of any run of three pictures, you will find this sum underneath the prism right next to the third one of your run of three. Yet another one is that if you add any two consecutive pictures, you will find this sum on the third prism after the second one you have picked.

Try to find other features of the sequence.

8. How to Get All the Pictures into a Sequence

Why not try the following rule as shown in Table 23 (suggested by Roy Davies, of Leicester)?

Table 23

On present picture	On next picture
Yellow	Green
Red	Yellow
Add the green, the yellow and the red	Multiply the sum by T

To save you some headaches, Table 24 shows the first part of the sequence (one-third of it, to be exact):

Table 24

OO1	DO1	DD1
O1T	O1D	D1T
1T1	1DD	1TO
T1D	DDT	TO1
1DO	DTD	O11
DOD	TDD	11O
ODO	DDD	1OO

after which it starts again with OOT and the whole table repeats itself, but each picture is multiplied by T, in other words you are back to the first prism, but you have to roll it towards you to get the picture. When it has finished that part of the sequence, our rule makes it start again with OOD and run another round through the table. Each picture will then be a multiple by D of the corresponding picture of the first round.

If you add any two consecutive pictures, you will get the one level with the second one in the next column on the right. If you try this with pictures in the third column, you will get a picture in the first column of the second round, again at the same level as your second picture that you picked.

It might be of interest to see where the families are in this sequence. The easiest ones to spot are the ones with a colour missing, as you just have to look out for the zeros. If we write Y for belonging to a family and N for not belonging to a family you can observe the following regular pattern:

Y N N Y Y N N N N Y N Y N N N N N N N N

You can also observe the same pattern in the case of the other families. The equal ones are easy to spot. The eleventh picture is DDT and all the other pictures in which the green is equal to the yellow follow the suggested pattern, in the order:
DDT — — DDD DD1 — — — — 11O – OOT — — — — — — — — — —
There are many ways you can make up sequences which take up all the pictures. The rules for getting the next green and the next yellow can stay the same as in the rules given above. There are twelve different ways of making up the rule for getting the red part of the next picture. Here they are, starting with the rule we already have:

(T × green) + (T × yellow) + (T × red) = red next
(T × green) + (T × yellow) + (D × red) = red next
(T × green) + yellow + red = red next
(T × green) + yellow + (D × red) = red next
(T × green) + (D × yellow) + (T × red) = red next
(T × green) + (D × yellow) + red = red next
....
(D × green) + (T × yellow) + red = red next
(D × green) + (T × yellow) + (D × red) = red next
(D × green) + yellow + (T × red) = red next
(D × green) + yellow + red = red next
(D × green) + (D × yellow) + (T × red) = red next
(D × green) + (D × yellow) + (D × red) = red next

You might have noticed that the last six can be obtained from the first six by exchanging D and T. Doing this you will find another rule reflected in the dotted line above. Don't forget that 'multiplied by T' means taking the one behind of the colour in question, and 'D times-ing' means taking the one underneath of the colour in question.

There are, of course, many other ways of getting sequences of 63 pictures by using rules that are different from our green next and yellow next rules. One easy way would be to take every other picture in the sequence. Since 63 is odd, you would go through all the pictures after two rounds. Of course you would have to find the rules that generate that sequence! These rules:

green next = (green + yellow + red) × T
yellow next = green × D + yellow + red
red next = green × T + yellow + red × D

would give you every third picture and so would use up only seven discs, and so would give you a sequence of 21 pictures. But if you tried

green next = red
yellow next = (green + yellow + red) × T
red next = green × D + yellow + red

you will get all the 63 pictures into your sequence.

You might check how the families are distributed in the various full sequences. You will find that some will give you the same pattern but backwards. Do you find any other different patterns? If so, what are they?

9. Different Ways of Laying Out the Families

Here are some fun ways of playing with your prisms. I hope you enjoy them.

A 'slide' will mean passing from one hexagon to another in the space between two adjacent hexagons. The two hexagons not in the big one will be called the extras.

Families should be placed in the following positions:

(i) Lines of little hexagons joining opposite vertices of the large hexagon.

(ii) Two adjacent sides between vertically shaded little hexagons.

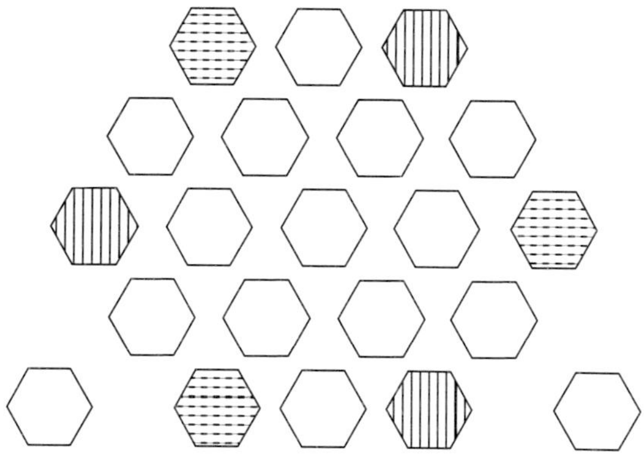

Figure 67

Start from a vertically shaded vertex, do a slide, then turn slightly and use the three hexagons making a line parallel to the side nearest to you. Add one of the extras.

Start from a horizontally shaded vertex, do a slide, then turn slightly and use two of three hexagons making a line parallel to the side nearest to you, by skipping over the middle one. Go back to start and do this again using another slide. You will get a pattern like a drinking cup.

Start from a horizontally shaded vertex, then go to the middle of each of the adjacent sides. Imagine a line joining these two middles, then use the two hexagons just slightly above this line (as seen from your starting point) adjacent to these middles. With a little imagination, you will see the shape of a lampshade!

The middles of non-adjacent sides, together with the centre and an extra.

The three horizontally shaded vertices, with both extras.

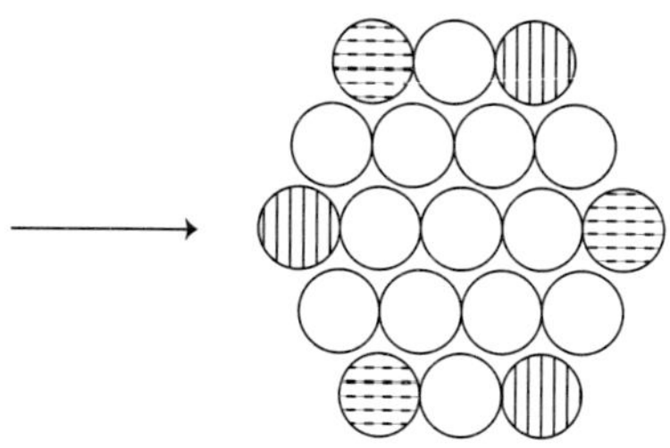

Figure 68

Figure 69 shows some examples of how the described patterns will look like on such a figure. The extras will not be shown.

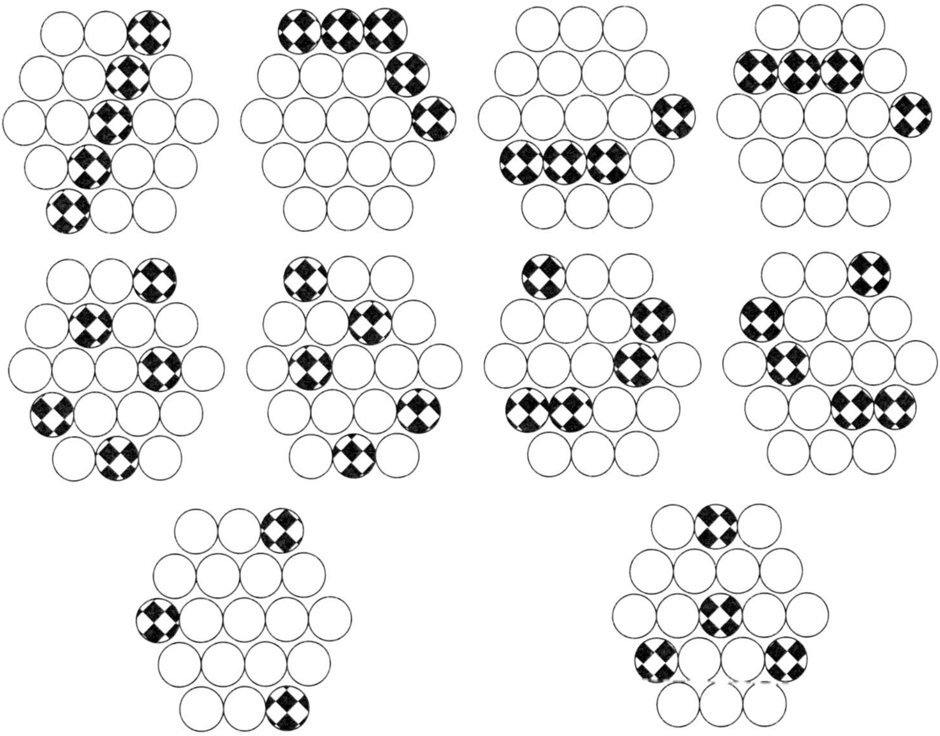

Figure 69

The patterns shown in Figure 69 are not necessarily compatible with each other. They are given so as to give the idea of the shape each pattern takes.

You have to make sure that you have the correct starting point, namely you must be aware of whether you are starting the pattern from a vertically shaded vertex or from a horizontally shaded vertex. You should make sure that the three families you choose for the three diagonals do have a common prism, as this will have to go in the centre of the diagram.

The diagram in Figure 70 is to help you place the families. The dotted arrows only pick out four circles; the fifth member of each family should be placed in one of the extra circles.

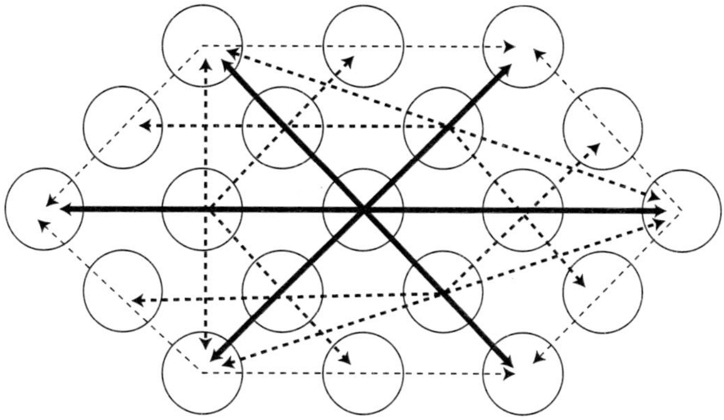

Figure 70

10. Mathematical Notes for Those Who Might Be Interested

A list of companion matrices in the three-dimensional vector space with the four-element Galois field as basis field is found below, together with their orders. (The order is the maximum number of times that you need to apply the matrix consecutively before you get back to the first vector to which you have applied it.)

A companion matrix in this vector space is of the following type:

0 1 0
0 0 1
a b c

where a, b and c are elements of the four element field. I shall refer to each matrix by writing the third row only, since the first two rows are the same in all the matrices to be considered.

I shall use the digits 0, 1, 2 and 3 for the elements. For convenience here are the addition and multiplication tables of the field (see Figure 71).

+	0	1	2	3
0	0	1	2	3
1	1	0	3	2
2	2	3	0	1
3	3	2	1	0

*	0	1	2	3
0	0	0	0	0
1	0	1	2	3
2	0	2	3	1
3	0	3	1	2

Figure 71

Order 3: [1 0 0]
Order 4: [1 1 1]
Order 5: [1 2 2], [1 3 3]
Order 6: [2 2 1], [3 3 1], [2 1 2], [3 1 3], [2 3 3], [3 2 2]
Order 7: [1 1 0], [1 0 1]
Order 9: [2 0 0], [3 0 0]
Order 12: [1 2 3], [1 3 2]
Order 15: [1 2 1], [1 3 1], [1 1 2], [1 1 3], [2 0 2], [3 0 3], [2 0 1], [3 0 1], [2 0 3], [3 0 2], [2 2 0], [3 3 0], [2 1 0], [3 1 0], [2 3 0], [3 2 0]
Order 21: [1 0 2], [1 0 3], [1 2 0], [1 3 0]
Order 63: [2 2 2], [3 3 3], [2 2 3], [3 3 2], [2 1 1]. [3 1 1] [2 1 3], [3 1 2], [2 3 2], [3 2 3], [2 3 1], [3 2 1]

There are, of course, a lot of matrices that have the same order as each one of the above companion matrices. All you have to do, if you want another matrix of the same order, is to pick a companion matrix of the order desired and find another one in the same conjugate class. You do this by choosing any non-singular matrix in the vector space, say the matrix A, C being the companion matrix chosen, and calculate the matrix

A ⋆ C ⋆ (INV)A

where (INV)A is the inverse matrix of A.

I believe all matrices of order 63 have the two-dimensional subspaces embedded in the sequence of vectors generated by them in the following pattern or in the same pattern in the reverse order:

Y N N Y Y N N N N Y N Y N N N N N N N N N

the pattern repeating twice more. The Ys mean that the vector belongs to a subspace and the Ns that it does not.

I also believe that you get the forward pattern if the elements in the third row of the matrix are either all equal or all different, otherwise you get the same pattern but backwards. But I have not proved this.

We have been looking at the problems from the point of view of three dimensions. There is no reason why we should not regard all objects on the faces of our prisms as separate elements. In other words, instead of having three colours, and having either nothing or a tree or a star or both in each colour, we could consider the trees and the stars as separate elements, even when they are in a double. This results in going to six dimensions from three, but in each dimension there would be only two values: *is* or *isn't*, in other words 1 or 0. The rule of succession of the sequence at the beginning is shown in Table 25.

Table 25

Picture you are looking at (vector)	Next picture	(Next vector)
Yellow star	Green star	0 0 1 0 0 0
Yellow tree	Green tree	0 0 0 1 0 0
Red star	Yellow star	0 0 0 0 1 0
Red tree	Yellow tree	0 0 0 0 0 1
Odd number of trees in total	Red star	0 1 0 1 0 1
Odd number of trees and stars in total	Red tree	1 1 1 1 1 1

The matrix is given above on the right. The variables come in the following order:

green star, green tree, yellow star, yellow tree, red star, red tree

11. Postscript

Figure 72 shows the spaces on which prisms should be placed which belong to the same family. Not all the families are included as doing so would make the diagram rather more confusing than clarifying! The situation of each family is shown by means of different arrows. It will be seen that some arrows only join four spaces together. The fifth member of the family is then always one of the extras.

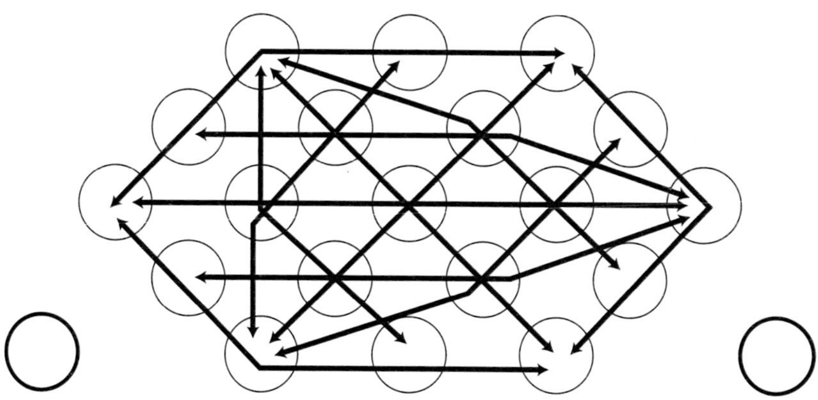

Figure 72

It is best to start in the middle, either with the prisms with everything equal on each face, or with one of the two prisms on which everything is different on each face. You will then get your prisms symmetrically arranged.

Chapter XIII
SEQUEL TO DESARGUES AND CROSS-AXIS

1. Enlargement of the Tennis Party Problem

You might have wondered whether it was possible to extend the tennis parties of the first Desargues game to larger teams. Here is one such extension:

Instead of 13 players, let us now have 21 players. The game can still be tennis, but the teams should now have five members, one of whom is elected to be the umpire. There will be super-umpires later on, in cases when the players strongly disagree with the normal umpire's decisions.

The problem again is to make teams, but now teams of five players, so that everyone plays with every other player (being umpire counts as playing) in the same team, but only once. The players are shown in Table 26.

Table 26

Adam	Betsy	Carol	Dick	Eric	Fred	George
Harry	Ian	John	Kevin	Laurie	Michael	Nelly
Oliver	Peter	Wendy	Rupert	Steve	Tom	Ursula

Betsy wants to be in the team in which Fred and John play, as well as in another in which George and Kevin play, and finally in yet another team in which Harry and Laurie play. For convenience these are written in the same columns.

If you find difficulty in setting up the 21 groups of five players, Table 27 shows one solution.

Table 27

Adam, Betsy, Carol, Dick, Eric	Adam, Fred, George, Harry, Ian
Adam, John, Kevin, Laurie, Michael	Adam, Nelly, Oliver, Peter, Wendy
Adam, Rupert, Steve, Tom, Ursula	Betsy, Fred, John, Nelly, Rupert
Betsy, George, Kevin, Oliver, Steve	Betsy, Harry, Laurie, Peter, Tom
Betsy, Ian, Michael, Wendy, Ursula	Carol, Fred, Kevin, Wendy, Tom
Carol, Ian, Laurie, Nelly, Steve	Carol, Ursula, Peter, George, John
Carol, Rupert, Oliver, Harry, Michael	Dick, John, Harry, Wendy, Steve
Dick, Tom, Nelly, George, Michael	Dick, Rupert, Kevin, Ian, Peter
Dick, Fred, Oliver, Ursula, Laurie	Eric, Fred, Steve, Peter, Michael
Eric, Ian, John, Oliver, Tom	Eric, Ursula, Nelly, Kevin, Harry
Eric, Rupert, George, Laurie, Wendy	

2. Method of Solution

If you are curious to know how I pulled these teams out of my hat, here is the trick. Place sixteen of the names in four rows of four as shown in Table 28.

Table 28

Fred	George	Harry	Ian
John	Kevin	Laurie	Michael
Nelly	Oliver	Peter	Wendy
Rupert	Steve	Tom	Ursula

The five players not mentioned are Adam, Betsy, Carol, Dick and Eric. They will form a team of five, the A–B–C–D–E team. Then Adam will complete the teams of the players whose names are in the rows, and Betsy will complete the teams from the columns. The other three players of the A–B–C–D–E team will complete the remaining teams in the following ways:

Take a diagonal such as Rupert, Oliver, Laurie, Ian. Turn the names of the first two through a right angle on the figure. You will get Nelly, Steve, Laurie, Ian. Taking the same diagonal again, turn the last two names through a right angle. You will get Rupert, Oliver, Harry, Michael. Then do the same with the other diagonal. You will get the four sets of four players each, each being completed by Carol.

Now take a sub-diagonal and the opposite corner. For example, John, Oliver, Tom, Ian. You can do this in four ways. Each of these groups of four will be completed into a team by Eric. The hardest one to explain is how Dick finds the players he is to play with. Dick is fond of parallelograms, so he takes, for example, John, Harry, Wendy, Steve, and another one like it (Nelly, George, Michael, Tom). Then he goes for some elongated parallelograms taking in two opposite vertices, such as Rupert, Kevin, Ian, Peter and Fred, Oliver, Ursula, Laurie.

3. Constructing the Cross-Axis

Since the teams have been picked for the situation when the quarrels between certain players have been settled, we have a suitable example for constructing the cross-axis. Betsy is the boss here and so it must be arranged that Fred and John play with Betsy, Kevin and George play with Betsy, Laurie and Harry play with Betsy, while Fred, George and Harry are in one and the same team and John, Kevin and Laurie are also in one and the same team. The cross-axis is obtained by identifying the following three players:

(i) The one who plays in the Fred–Kevin team as well as in the John–George team.

(ii) The one who plays in the George–Laurie team as well as in the Kevin–Harry team.

(iii) The one who plays in the John–Harry team as well as in the Fred–Laurie team.

Notice that the three possible crosses have been used for identifying the required players. For your convenience, the teams are reproduced in Table 29 so you can check that what is being suggested is correct.

Table 29

Adam, Betsy, Carol, Dick, Eric	Adam, Fred, George, Harry, Ian
Adam, John, Kevin, Laurie, Michael	Adam, Nelly, Oliver, Peter, Wendy
Adam, Rupert, Steve, Tom, Ursula	Betsy, Fred, John, Nelly, Rupert
Betsy, George, Kevin, Oliver, Steve	Betsy, Harry, Laurie, Peter, Tom
Betsy, Ian, Michael, Wendy, Ursula	Carol, Fred, Kevin, Wendy, Tom
Carol, Ian, Laurie, Nelly, Steve	Carol, Ursula, Peter, George, John
Carol, Rupert, Oliver, Harry, Michael	Dick, John, Harry, Wendy, Steve
Dick, Tom, Nelly, George, Michael	Dick, Rupert, Kevin, Ian, Peter
Dick, Fred, Oliver, Ursula, Laurie	Eric, Fred, Steve, Peter, Michael
Eric, Ian, John, Oliver, Tom	Eric, Ursula, Nelly, Kevin, Harry
Eric Rupert, George, Laurie, Wendy	

It is easily checked that the person under (i) is Carol, the one under (ii) is Eric, and the one under (iii) is Dick. These three players in fact play in the very first team we constructed. So the cross-axis property is seen to work in the 21-player problem, as well as in the 13-player problem.

4. Using the Enlarged Situation for Desargues's Theorem

The two 'quarrelsome' threesomes could be, for example, Fred, John, Oliver, and Harry, Laurie, Peter, the pairs Fred and Harry, John and Laurie, and Oliver and Peter all playing with Adam in three different teams, so the conditions of Desargues's theorem are satisfied. So we now have to look for the following players:

(i) The person who plays in the Fred–John team as well as in the Harry–Laurie team.
(ii) The person who plays in the John–Oliver team as well as in the Laurie–Peter team.
(iii) The person who plays in the Fred–Oliver team as well as in the Harry–Peter team.

It is not hard to see that (i) is Betsy, (ii) is Tom and (iii) is Laurie by looking through the list of teams. Betsy, Tom and Laurie play in one and the same team, Peter and Harry being the other two members of their team.

You could try some more quarrelsome threesomes (in geometry – triangles!) You will find that Desargues's theorem is verified in each case.

Table 30

Adam	Fred	George	Harry	Ian	Betsy	1 → 2 → 3
xyz	xxx	yx0	0yx	x0y	xzy	
Adam	John	Kevin	Laurie	Michael	John	2 → 1
xyz	xy0	00x	xyx	xyy	xy0	
Adam	Nelly	Oliver	Peter	Wendy	Nelly	3 → 2
xyz	0xy	yxy	x00	xxy	0xy	
Adam	Rupert	Steve	Tom	Ursula	Rupert	1 → 3
xyz	y0x	yxx	yyx	0x0	y0x	
Betsy	Fred	John	Nelly	Rupert	Adam	2 → 1 → 3
xzy	xxx	xy0	0xy	y0x	xyz	
Betsy	George	Kevin	Oliver	Steve	George	1 → 2
xzy	yx0	00x	yxy	yxx	yx0	
Betsy	Harry	Laurie	Peter	Tom	Harry	2 → 3
xzy	0yx	xyx	x00	yyx	0yx	
Betsy	Ian	Michael	Wendy	Ursula	Ian	3 → 1
xzy	x0y	xyy	xxy	0x0	x0y	
Carol	Fred	Kevin	Tom	Wendy	Carol	1 = 2
xx0	xxx	00x	yyx	xxy	xx0	
Carol	Ian	Laurie	Nelly	Steve	Wendy	3 → 1 2
xx0	x0y	xyx	0xy	yxx	xxy	
Carol	Ursula	Peter	John	George	Kevin	no 3
xx0	0x0	x00	xy0	yx0	00x	
Carol	Rupert	Oliver	Harry	Michael	Tom	1 2 → 3
xx0	y0x	yxy	0yx	xyy	yyx	
Eric	Fred	Steve	Peter	Michael	Eric	2 = 3
0xx	xxx	yxx	x00	xyy	0xx	
Eric	Ian	John	Oliver	Tom	Michael	2 3 → 1
0xx	x0y	xy0	yxy	yyx	xyy	
Eric	Ursula	Nelly	Kevin	Harry	Peter	no 1
0xx	0x0	0xy	00x	0yx	x00	
Eric	Rupert	George	Laurie	Wendy	Steve	1 → 2 3
0xx	y0x	yx0	xyx	xxy	yxx	
Dick	Fred	Laurie	Ursula	Oliver	Dick	1 = 3
x0x	x0x	xyx	0x0	yxy	x0x	
Dick	Ian	Peter	Rupert	Kevin	Ursula	no 2
x0x	x0y	x00	y0x	00x	0x0	
Dick	Nelly	George	Michael	Tom	Oliver	1 3 → 2
x0x	0xy	yx0	xyy	yyx	yxy	
Dick	John	Harry	Wendy	Steve	Laurie	2 → 1 3
x0x	xy0	0yx	xxy	yxx	xyx	
Adam	Betsy	Carol	Eric	Dick	Fred	add as is
xyz	xzy	xx0	0xx	x0x	xxx	

5. Suitable Coding System for the Players and the Teams

Table 30 shows the teams of five, along with the super-umpires. A suggested code is given under each player and a code is given for each team. You might like to work out the rationale for the coding system, but if you read on you will find some explanations. The xyz codes will help you find out how I have picked the super-umpires. The numerical code at the end of each team will help you to know the reason why the members of those teams belong together.

6. Some Algebraic Explanations

Here are some interesting aspects of the 21 elements expressed in codes and the 21 teams that I used for solving the 21-person tennis party problem.

First a little mini-algebra, if you can stomach it! In our mini-algebra we have four elements: 0, x, y and z. The element 0 behaves just like 0 in adding.

Adding an element to itself, the sum is 0, so:

$$x + x = y + y = z + z = 0$$

Adding any two different elements out of x, y, z, the sum is always the third element. You can add elements in either order; it will not make a difference to the sum.

Each player code has three of the above elements, given in a certain order. Each such code can be replaced by another code, representing the same player, by changing x into y, y into z and z into x, but leaving the 0 unchanged as 0. So each player has three codes. I have tried to use x and y as much as possible; only two codes have three different non-0 elements in them: they are Adam's and Betsy's codes. So by using the cyclic change described above, I have been able to eliminate all other z's from the other codes.

Now try to add any two codes belonging to the same team, adding the first letter to the first letter, the second letter to the second letter and the third letter to the third letter. You will usually have to do some cyclic changes to get rid of any z's, but you will find that you always obtain a code belonging to the same team.

Take, for example, Michael and Wendy who play in Ian's team:

Michael + Wendy = x y y + x x y = 0 z 0 = 0 x 0 (by cyclic change) = Ursula

And we can see that Ursula is also in Ian's team.

Now take a look at the properties given for the teams. The numbers refer to the positions of the elements. For example, no 3 means that the third element is always 0; 2 = 3 means that the second element is equal to the third element and so on.

Now let me explain the arrows. The element or elements at the pointed end of the arrow must move up one notch in the cyclic change. Having done this move, look at the elements referred by numbers. These are the relevant elements. Check whether the sum of the relevant elements is 0. If so, the code possesses the property in question, if not, it does not possess the property in question.

In the cases where there are two arrows, the element at the end of the first arrow moves up one notch, and the element at the end of the second arrow, moves up two notches in the cyclic order

x → y → z → x

Let me illustrate in Table 31 by doing a few examples. Take Harry's team.

Table 31

Betsy	Harry	Laurie	Peter	Tom	Harry	2 → 3
x z y	0 y x	x y x	x 0 0	y y x	0 y x	

The property in question is 2 → 3. This means that only the second and the third elements are relevant. Take Betsy's code, namely x z y. If we move up the third element by one notch, we get x z z. Adding the second and the third elements gives us z + z = 0. Note that we ignore the first element, as the first element is not relevant, since it is not mentioned in the code 2 → 3.

So the property is verified for Betsy's code. If we do the same to the other codes, we obtain 0 y y for Harry, x y y for Laurie, x 0 0 for Peter and y y y for Tom, not forgetting that moving up 0 does not move it, but leaves it as it is.

So the second element and the third moved up element for all the members of the team always add up to 0 by our rules.

Now let us look at Wendy's team as shown in Table 32.

Table 32

Carol	Ian	Laurie	Nelly	Steve	Wendy	3 → 1 2
x x 0	x 0 y	x y x	0 x y	y x x	x x y	

The corresponding property is 3 → 1 2. This means that the elements at the first and at the second places have to be moved up one notch, and the sum of all the positions so obtained must give 0. Here all three elements are relevant. The third one stays as it is and the other two each move up one notch. If we carry this out, we shall get y y 0 for Carol, y 0 y for Ian, y z x for Laurie, 0 y y for Nelly and z y x for Steve, and you can check that they all come to 0 when added.

Finally, let me do a double arrow, taking Betsy's team as shown in Table 33.

Table 33

Adam	Fred	George	Harry	Ian	Betsy	1 → 2 → 3
x y z	x x x	y x 0	0 y x	x 0 y	x z y	

Betsy' s team has the property 1 → 2 → 3. This means that the second element must be moved up by one and the third element by two notches in the cyclic change. If we do this, we get x z y for Adam, x y z for Fred, y y 0 for George, 0 z z for Harry and x 0 x for Ian, and if you add the elements, in all five case you get the sum 0.

No doubt you will have checked by now that any three teams with a player in common have super-umpires who belong to one and the same team. You might also have noticed that the super-umpire of this team is always the player who is common to the three teams examined.

7. Three in a Team or Six in a Team

I almost forgot. How about seven people wanting to play triangular tennis? Table 34 shows the teams of three with the umpires given.

Table 34

Ann, Betsy, Carol	Umpire is Fiona
Betsy, Evelyn, Gladys	Umpire is Debra
Ann, Debra, Evelyn	Umpire is Evelyn
Carol, Debra, Gladys	Umpire is Gladys
Ann, Fiona, Gladys	Umpire is Carol
Carol, Evelyn, Fiona	Umpire is Ann
Betsy, Debra, Fiona	Umpire is Betsy

Table 35 shows the teams for 31 players wanting to play a game that takes 6 players each time. Every player is playing only once with every other player. The umpire problem is left for you to sort out!

Table 35

Ann, Betsy, Carol, Debra, Evelyn, Fiona
Betsy, Adam, Fred, Kevin, Quixote, Vic
Betsy, Bill, George, Larry, Robert, William
Betsy, Charles, Harry, Michael, Steve, Xerxes
Betsy, Dick, Ian, Norman, Tom, Yvan
Betsy, Eric, John, Peter, Urquhart, Zoltan
Debra, Eric, George, Kevin, Steve, Yvan
Debra, Adam, John, Norman, Robert, Xerxes
Debra, Charles, Ian, Larry, Quixote, Zoltan
Debra, Bill, Harry, Peter, Tom, Vic
Debra, Dick, Fred, Michael, Urquhart, William
Fiona, Charles, George, Norman, Urquhart, Vic
Fiona, Dick, Harry, Kevin, Robert, Zoltan
Fiona, Adam, Ian, Peter, Steve, William
Fiona, Eric, Fred, Larry, Tom, Xerxes
Fiona, Bill, John, Michael, Quixote, Yvan

Ann, Adam, Bill, Charles, Dick, Eric
Ann, Fred, George, Harry, Ian, John
Ann, Kevin, Larry, Michael, Norman, Peter
Ann, Quixote, Robert, Steve, Tom, Urquhart
Ann, Vic, Williams, Xerxes, Yvan, Zoltan
Carol, Adam, George, Michael, Tom, Zoltan
Carol, Charles, Fred, Peter, Robert, Yvan
Carol, Eric, Harry, Norman, Quixote, William
Carol, Bill, Ian, Kevin, Urquhart, Xerxes
Carol, Dick, John, Larry, Steve, Vic
Evelyn, Dick, George, Peter, Quixote, Xerxes
Evelyn, Charles, John, Kevin, Tom, William
Evelyn, Bill, Fred, Norman, Steve, Zoltan
Evelyn, Adam, Harry, Larry, Urquhart, Yvan
Evelyn, Eric, Ian, Michael, Robert, Vic

Check Desargues's theorem and the cross-axis with the above teams!

Let me give a few little algebraic secrets away. Let N be the number one less than the number of people needed to make up a game. Then it is easy to see that the number of players and the number of games played is:

$$N^2 + N + 1$$

If there is just one player playing patience, the formula gives N = 0 and the number of games is, of course, 1. If there are two people playing each of the games, then N = 1, and the number of players as well as the number of games to be played is 3. For three people playing a game, N = 2, and our formula gives 7 players and 7 games. For the foursomes, we need 13 people, since then N = 3, and the formula gives $3^2 + 3 + 1 = 13$. For the five-person tennis, we have N = 4, and so the number of people and of games is given by $4^2 + 4 + 1 = 21$, and for the six-person games N = 5, and the number of players and the number of games is given by $5^2 + 5 + 1 = 31$.

I shall leave it to you to ponder why this magic formula works and on the problem of whether it would still work for a seven-person set of games and if so, why, and if not, why not?

Chapter XIV

HOW TO PLAY WITH FIVE IMAGINARY PEOPLE, OR WITH FIVE PATTERNS

1. Introduction to the two Games

Imagine five people: Oliver, William, Alice, Bruce and Carol. I shall refer to them for brevity's sake sometimes as 0, 1, A, B, C. There are two games these people can play. Each game is played either by just two people or by a single person. One game is called the S game, the other one is called the P game. These imaginary people agree on a rule about who should play next. When two of them (or maybe just one of them) have been playing a game, then there is a rule to decide who should play next. This rule is different after an S game from what it is after a P game. The person who is to play next as determined by the rule then decides with whom he or she wants to play next, whether to play patience, and whether to play the S game or the P game. For the S game the players who have been playing have to consider the cycle

$$1 \rightarrow A \rightarrow B \rightarrow C \rightarrow 1$$

and use the following rules:

(i) If a person has been playing patience, then the next person in the above cycle will play next.
(ii) If two 'neighbouring' players in the cycle have been playing, then the player before the first one of the two in the cycle will play next.
(iii) If two players separated in the cycle by a person have been playing, then Oliver will play next.
(iv) If Oliver has been playing with another player, then this other player will play next.
(v) If Oliver has been playing patience, then Oliver will play next.

For example, S BB = C, S BC = A, S AC = 0 and so on, meaning that if Bruce has been playing S game patience, then Carol will be one of the next players, if Bruce and Carol have been playing the S game, then Alice will play next, and if Alice and Carol have been playing the S game, then Oliver will be one of the next players. Oliver in the S game behaves like 0 in adding.

For the P game the same cycle is used but with different rules. These are the P rules:

(i) If Oliver has been one of the players, then Oliver will play again.

(ii) If William has been one of the players, then the one he played with will play next.

(iii) If Alice has been one of the players, then the next in the cycle to the player with whom Alice has played will play next. Alice replaces her partner by the next person to her partner in the cycle. Of course, this person may or may not choose Alice as partner!

(iv) If Bruce has been one of the players, then the next but one in the cycle to the one Bruce has played with will play next. Bruce replaces his partner by the person two notches up in the cycle. Again, this new player may or may not choose Bruce as partner!

(iv) If Carol has been one of the players, then the previous person in the cycle to the one with whom Carol has played will play next. Carol replaces her partner by the person just before this partner in the cycle, who will then choose with whom he or she will play.

Oliver is the greedy player. Once he has got into the P game, there is no way to stop him, except by choosing to play the S game. On the other hand William is very generous, always letting his partner in the game play next.

If you have trouble remembering these things, have these in front of you on a piece of paper while you are working this mini-algebra:

P AA = B, P BB = 1, P CC = B, P AB = C, P BC = A, P AC = 1

You might have noticed that Oliver acts like 0 in multiplication when playing the P game, and William acts like the 1 in multiplication in the P game. So the S rules give you the sum of the two players who have been playing and the P rules give you the product of the two players who have been playing.

So instead of using the symbols S and P, we could use adding and multiplying signs to show which of the two games the people were playing!

2. How to Play with Real People

Practise these moves in the S game and in the P game before you go on. It is easy to make a mistake and get the wrong answer. You could choose five people: one could be Oliver, one could be William, one could be Alice, one could be Bruce and one could be Carol. You could arrange five chairs facing towards each other, and arrange yourselves so that if anyone walked round the chairs in a clockwise sense, he or she would meet the players in this order: Oliver, William, Alice, Bruce, Carol.

The S game could be that the two players would whirl each other round linking arms, and the P game could be the same but you would hold each other by both hands.

If you do not feel so energetic, you could use patterned cubes. You could choose five patterns instead of five people. For example you might make the following choices:

Oliver = black, William = white, Alice = criss-cross, Bruce = level, Carol = shaded

In what follows I shall use mainly these patterns, but I might refer to the players as well, calling them 0, 1, A, B and C respectively.

3. How to Play with Patterned Cubes – Operations with Towers of Cubes

We can make towers of three with our cubes. Since there are five patterns, it is possible to make 125 such towers!

Let us now play with these towers. The obvious way of using the S rules for two towers is to use the rule for the first cube of each tower then use the rule for the second cube of each tower and finally for the third cube of each of the two towers. In this way any two towers will result in another tower.

We can also use the P rules in the case of a tower and just one of our cubes simply by using the rule as between each cube of the tower and the cube we selected. This operation is often referred to as the P operating of a tower by a cube or cube to tower product.

I shall need another operation which is somewhat strange. Given two towers, this operation yields a cube. Given two towers, each consisting of a first, second and third cube, we can P operate the two first cubes, then the two second cubes and then the two third cubes and then S operate the three cubes so obtained. Such an operation is often referred to as cube P operation, or cube product (not to be confused with the previously defined P operating a tower by a cube).

Here are some examples of these three operations. First let me S operate two towers:

S (A1B, BB0) = 10B since S AB = 1, S 1B = 0 and S B0 = B

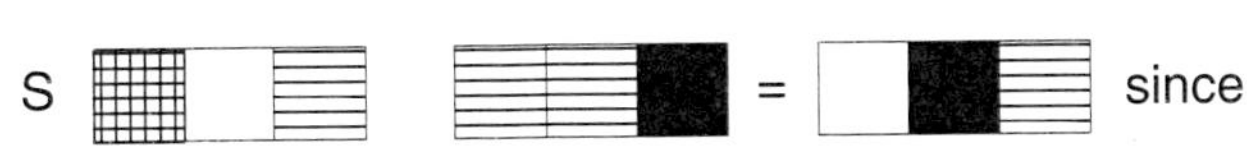

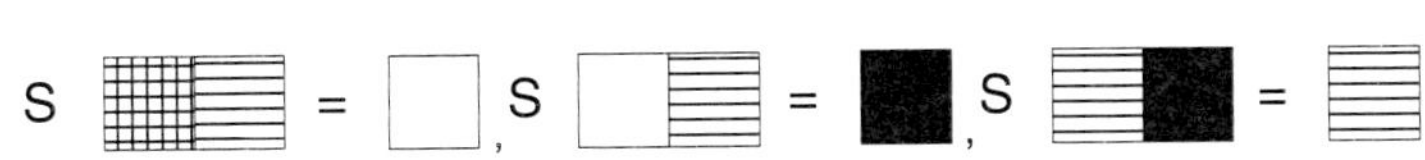

Now let me do a P operating of a tower by a cube
P A (A1B) = BAC since P AA = B, P A1 = A and P AB = C

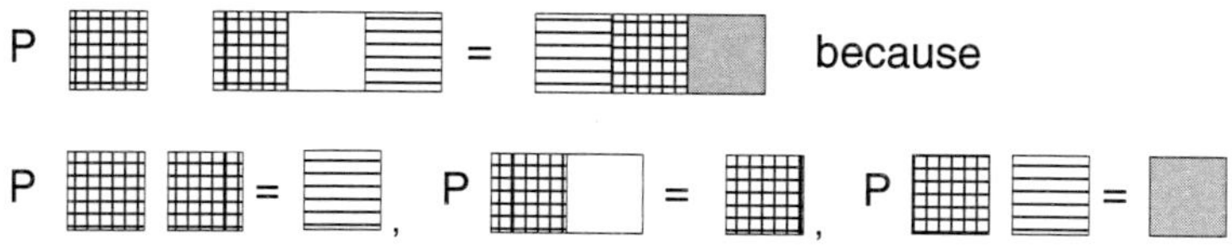

Now here is a cube P operation:
P (A1B, 0AA) = S (S (P A0 P 1A), P BA) = SAC = 0

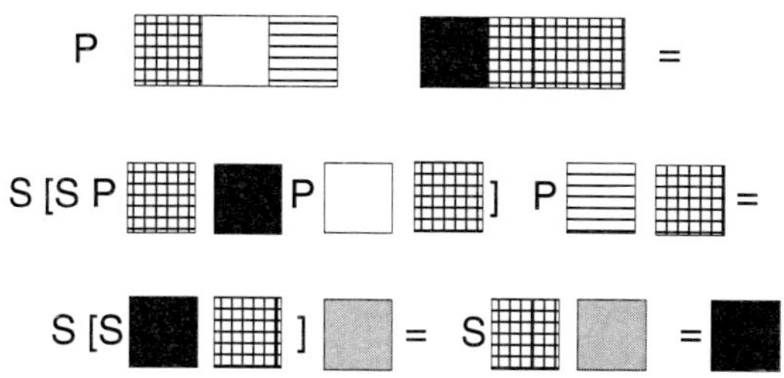

If the result of a cube P operation of two towers is 0, we sometimes say that the two towers are at right angles to one another. You should practise these three operations, as they will be used in what follows.

You might have been a little perplexed when I mixed up the P operation with the S operation, or even two successive S operations might have seemed somewhat confusing. To S operate on three cubes, what we have to do is to S operate on two of them and then S operate this result with the third cube. You can check that according to our rules it makes no difference which two cubes you work with first, combining the result with the third cube. In mathematical language we say in such cases that the S operation is associative. We can associate any two cubes, the result of which can then be associated with the remaining cube, and the resulting cube in the end will be the same one, whichever two we start with.

4. First Order Families of Towers

Choose any tower, except 000. Let us call it xyz. Out of this tower, you can generate other towers by P operating on it by a cube. Since there are five cubes, this would generate five different towers, including 000, which of course you can get by P operating xyz by the cube 000. Here is an example:

Suppose we choose ABC. Then P operating by the cubes 0, 1, A, B and C in turn, we get the five different towers:

000, ABC, BC1, C1A, 1AB

Now try S operating any two of the above towers. You will always get one of these five towers as the sum. For example, S[BC1 C1A] = ABC, and so on.

You can see that we are enclosed inside this small system of five towers. Such a sub-system is called a family.

The above family is of the first order, or of order 1, as it is obtained by starting out with one single tower. You will notice that all the towers of this family can be obtained by choosing any one of the towers besides 000, and cyclically permuting the people 1, A, B, C or the patterns white, criss-cross, level, grey.

In some of the problems I shall identify the four non-0 names of the elements in a family with each other, since all of them belonging together form the family. One of the members of the family will be selected to represent the whole family. So in fact, although such a family will have four names, as far as possible, only one of the four will be used to refer to the family in question.

5. Second Order Families

Choose any two towers, but not from the same first order family. Then start by writing down all the towers of the family generated by one tower, and then the towers of the family generated by the other tower. Then try to S operate all these towers together, thus generating more towers. You will know that it is a waste of time adding towers from the same first order family, because you will already have that tower in your list. So you should always choose one tower from one family to combine with a tower from the other family. You should get sixteen towers in all, including the tower 000.

Here is one way of doing it. Take these two towers as shown in Figure 73.

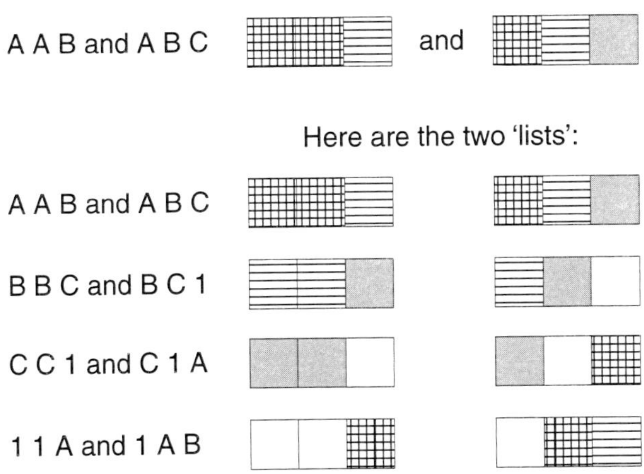

Figure 73

So S operating all to all in the two lists, we obtain these towers:

B1A, 100, 0C1, CBC, 1C1, CAB, A00, 01A, 0AB, A1A, 1BC, B00, C00, 0BC, BAB, AC1

The tower 000 we can also get, but only by taking two towers in the same list, but separated by another tower of the list.

So we can sort them into first order families like this, each arranged in cyclic order:

{000, 100, A00, B00, C00}
{000, A1A, BAB, CBC, 1C1}
{000, AAB, BBC, CC1, 11A}

{000, 01A, 0AB, 0BC, 0C1}
{000, AC1, B1A, CAB, 1BC}
{000, ABC, BC1, C1A, 1AB}

It is perhaps worth noticing that the third cube in each tower is always the one that comes after the second cube in the cyclic order. It is even more remarkable that no other tower has that property. In fact, every family has one property which is true of all its towers but not true of any other tower in our system. So you should now construct your own family and find its special property. Good hunting!

Calculate the cube product of any of the towers of the second family by any other tower in the set (even by itself). You will find that you always get 0. So the towers in the second family are at right angles to all other towers. They are even at right angles to themselves! I suppose at this point you might feel like rejecting the idea that a zero cube product can be interpreted as two towers at right angles to each other!

6. The Second Order Families and their First Order 'Friends'

Let us do a bit of organising. I shall choose a tower to be a representative tower of each first order family. Then I shall choose a tower to represent each second order family. This second chosen tower may or may not be a tower of the second order family that it represents. Here is a list of the 31 towers representing the 31 first order families:

001 010 100 111

101 011 110 11A 1A1 A11 1A B 1AC 1BC

10A 01A 1A0 11B 1B1 B11 A1B A1C B1C

10B 01B 1B0 11C 1C1 C11

10C 01C 1C0

To each of these representative towers will correspond a second order family, all towers in the family being at right angles to the representative tower. So there are 31 second order families as well as 31 first order families.

Each second order family has a unique property which all its towers have but no towers not in that family have. I have already given one example of such a family, with its unique property, as well as with the first order family which is at right angles to it. This can be done 31 times.

To describe a property I shall use the numbers 1, 2 and 3 to refer to the first, the second and the third cube in the towers of the family. The word 'no' followed by a number will mean that the cube of that number is always black in that family. So I shall use the names no 1, no 2 and no 3 for this purpose.

I shall write the equals sign between two numbers if the cubes at the places indicated always have the same pattern in the family in question. So I shall use the symbols (don't faint!) 2 = 3, 1 = 3 and 1 = 2.

If a cube in one position can be obtained by S operating on the two cubes in the two remaining positions, I shall also use the equals sign. So the symbols 1 = 23, 2 = 13 and 3 = 12 will be used for these cases.

I shall use the symbols *on*, *x* and *b* after a number signifying the position of a cube in the tower. The symbol *on* will mean that the position or positions must be moved on one space in the cycle 1 → A → B → C → 1. The symbol *x* will mean moving up two spaces in the cycle. The symbol *b* will mean moving back one space (or on three spaces) in the cycle. For example, 2 = 3*x* will mean that the second cube is always has the same pattern as the third cube moved up two in the cycle. Or 2 = 13*on* will mean that the second cube is always equal to the cube obtained by S operating on the first and the third cubes each moved up one space.

Some properties are expressed by stating that some combinations of the cubes, moved or not, must yield black. For example, 123 means that S operating all three cubes should yield black as the result. Or 1, 2*b*, 3*x* means that S operating on the first cube, the second cube moved back one space and the third cube moved up two spaces must always come to black.

You could now work out all the 31 second order families, find the unique property characterising each one, and attach the first order family (giving its representative tower) which is at right angles to the second order family in question. Or you could turn over the page and check that my solution to the problem is correct.

Table 36 shows a first second order family.

Table 36

Property	Family	Associated tower
No 1	001 010 / 011 01A / 01B 01C	100

In Table 37, the third and the fifth of the above towers complete the families of the rows; the second, the fourth and the sixth towers complete the families of the columns. The properties of the families are given at the start of each row and at the bottom of each column.

Table 37

	001	10B	10A	101	10C	
No 2	100	101	10A	10B	10C	010
1 = 2	110	111	11A	11B	11C	1B0
1 = 2x	1B0	1B1	B1C	B11	1BC	110
1 = 2on	1C0	1C1	A1B	A1C	A11	1C0
2 = 1on	1A0	1A1	C11	1AB	1AC	1A0
	No 3	1 = 3	3 = 1on	1 = 3x	1 = 3on	

	01A	1C1	11A	1AB	1BC	
2 = 3	100	111	A11	B11	C11	01B
2 = 13	1BC	10B	1A1	A1B	110	1B1
1, 2b, 3on	1AB	1C0	10A	11C	1B1	A1B
1, 2on, 3b	1C1	B1C	11B	1A0	10C	1AC
3 = 12	11A	1AC	1B0	101	A1C	11B
	3 =2on	13 = 2on	3 = 12on	1, 2on, 3x	1, 2x, 3b	

	A1C	1A1	11C	B1C	01C	
123	C11	10B	1B0	1C1	11C	111
1 =23	101	A11	1AB	110	B1C	B11
1 = 23on	1C0	11A	10C	B11	1A1	A11
2 = 3x	11B	1B1	A1B	1AC	100	011
23 = 1on	1BC	1A0	111	10A	A1C	C11
	1, 2b, 3x	2 = 13on	12 = 3on	1, 2x, 3on	2 = 3on	

I have put the families in rows and in columns partly for fun and partly for saving space.

If you work out the cube product of the associated tower by any other tower in the row or in the column to which it is associated, you will always get a black cube (or Oliver!) as a result. For example:

$$P\ (C11,\ 1BC) = S\ S[P(C1), P(1B)],\ P(1C) = S\ S(CB),\ C = S(AC) = 0$$

looking at the last row above. You can also check that the result of S operating on the second and the third cube is always equal to the first cube moved up one in the cycle, taking any tower from that last row. Do not forget that this property does not necessarily hold for the associated tower which is at right angles to the family being considered.

Perhaps the most surprising aspect of our construction is the following:

Take any three of our second order families which all have a first order family in common. The three associated towers to these second order families (representing three first order families) will always belong to one and the same second order family. What is more, the 'associated tower' of this family is always the tower representing the common first order family to the three selected second order families.

If you find the above too complicated, then let me formulate things a little differently and give some indication how to solve the following problem, based on the facts described:

There are 31 people who want to play Scrabble. They arrange themselves in parties of six, but in such a way that each player plays with every other player just once. For every party of six they choose a person who will keep the scores and make the decisions whether a word is allowed or not. This person can be one of the players in the party or a person not playing in that party.

They manage to choose the scorers in such a way that for any three parties that have a player in common, the corresponding three scorers play in one and the same team. What is more, the scorer of this last team of scorers is always the person who is common to the three parties in question. Now you try to arrange the Scrabble parties!

To solve the problem, all you have to do is to associate a player to each tower representing a first order family of our construction. The teams of six will be the six first order families (in the form of their representative towers) belonging to a second order family. The scorer will be the person you have associated to the tower which is at right angles to all the towers in the second order family in question.

If you want to amaze your friends, just give them the Scrabble team problem, and see how they make out. Keeping the algebra under the hat, just produce the solution. It should have quite a magical effect. Then it is up to you whether you let the algebraic cat out of the bag.

Of course, it goes without saying that you can play the Desargues game as well as the cross-axis game with the teams you so construct.

Chapter XV

GAMES WITH THIRTY-ONE BOXES

1. Introducing the Boxes

The boxes should be made of cardboard and have a square cross section. This means that two of the faces of the box should be square shaped; the others should be shaped in the form of longish rectangles, allowing space for drawing three stick men on each face, using the five different arm positions as shown in Figure 74.

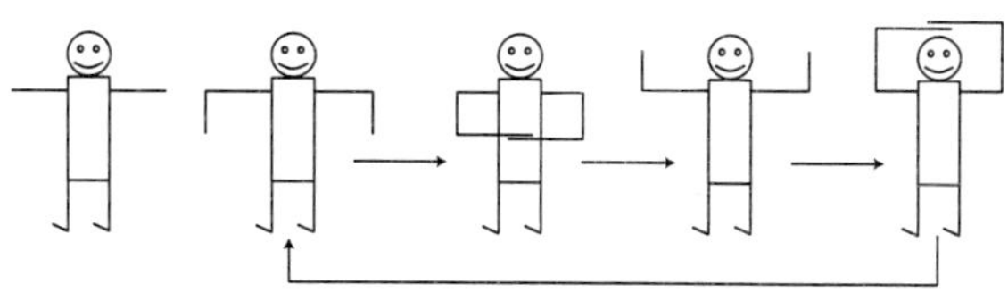

Figure 74

The position with the outstretched arms I shall call the zero position. The other positions are either upper or lower and either open or closed.

The arrows take you round these positions in this order: lower open, lower closed, upper open, upper closed and then back to lower open.

I shall define the sum of two positions as follows:

(i) Adding a zero to a position will leave that position unaltered.
(ii) Adding a position to itself will give the next position in the cycle.
(iii) Adding two positions next to each other in the cycle will give the position before the first of the two in the cycle.
(iv) Adding two positions with one in between them in the cycle will give the zero position.

I shall define multiplication of a position by another position as follows:

(i) Multiplying by zero will give the zero position.
(ii) Multiplying by upper open will leave the position unaltered.
(iii) Multiplying by upper closed will give the next position in the cycle.
(iv) Multiplying by lower open will simply invert the position to be multiplied.
(v) Multiplying by lower closed will give the previous position in the cycle.

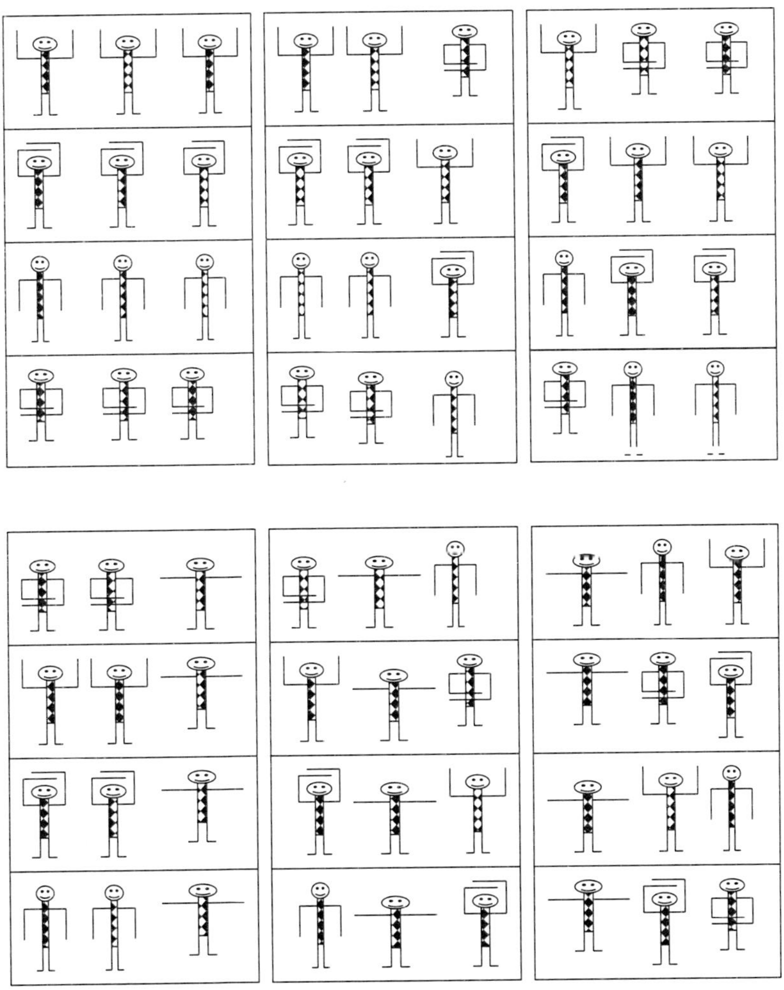

Figure 75

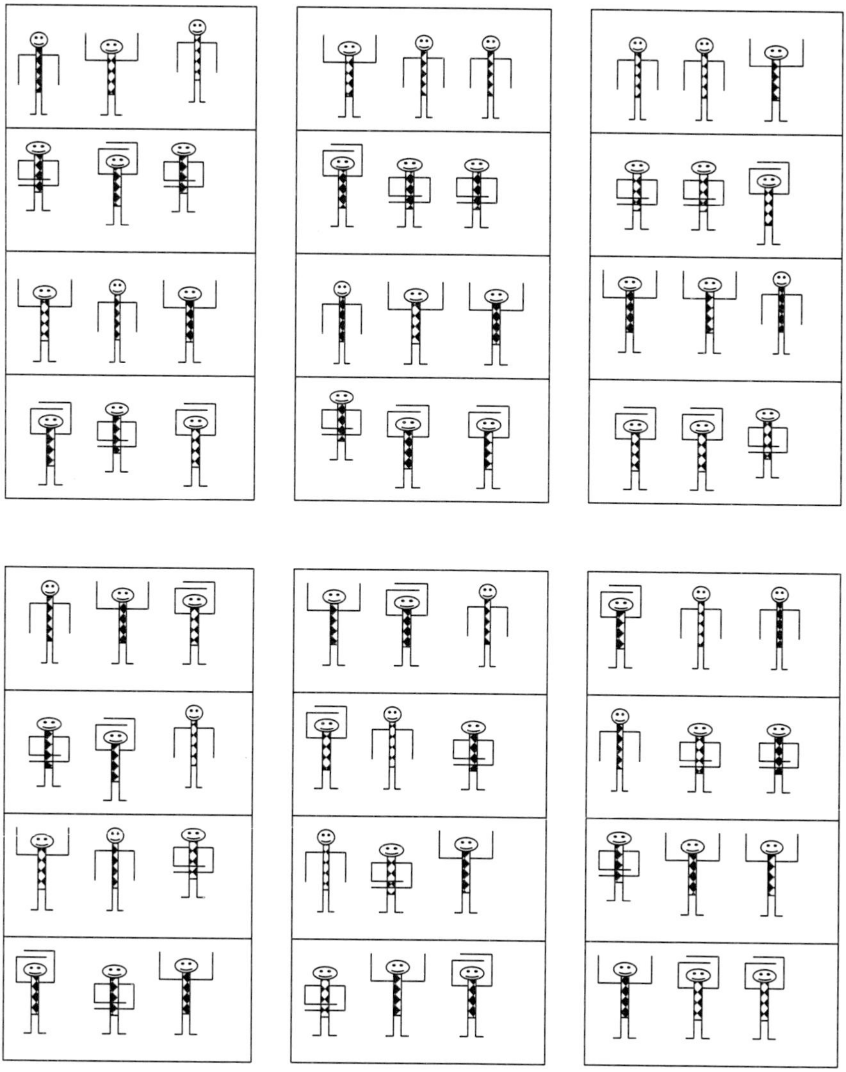

Figure 76

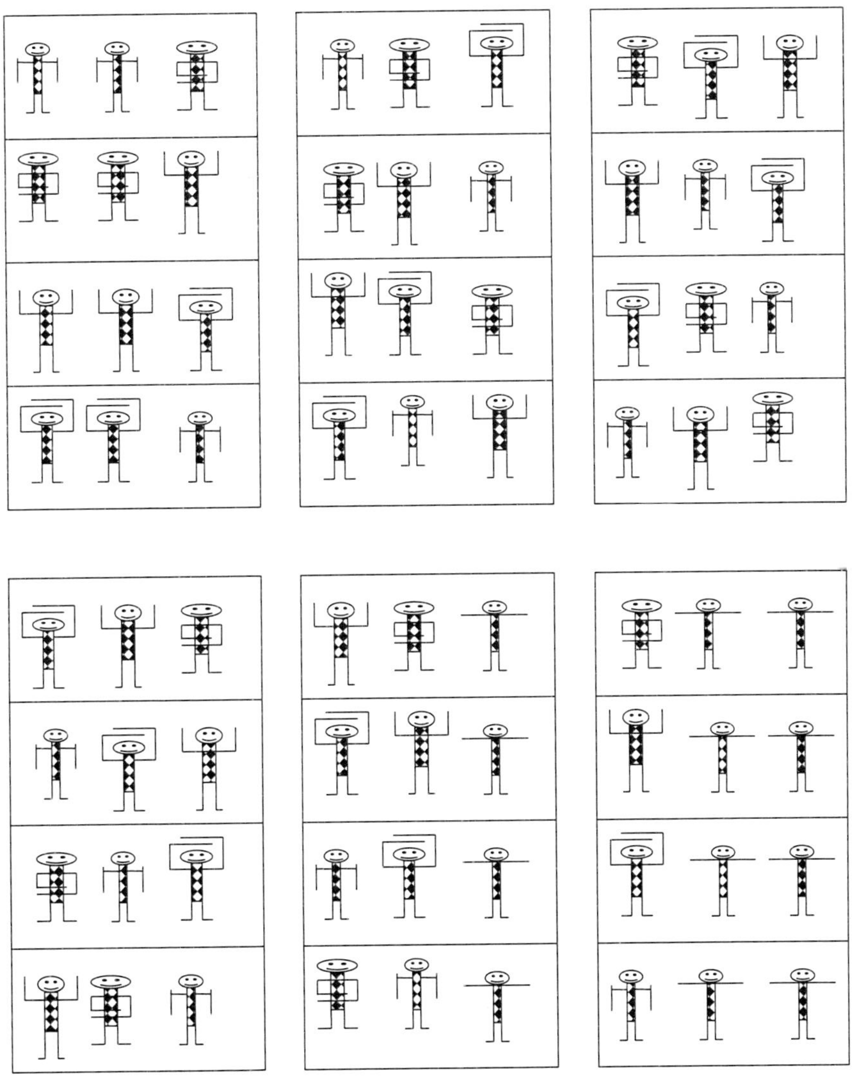

Figure 77

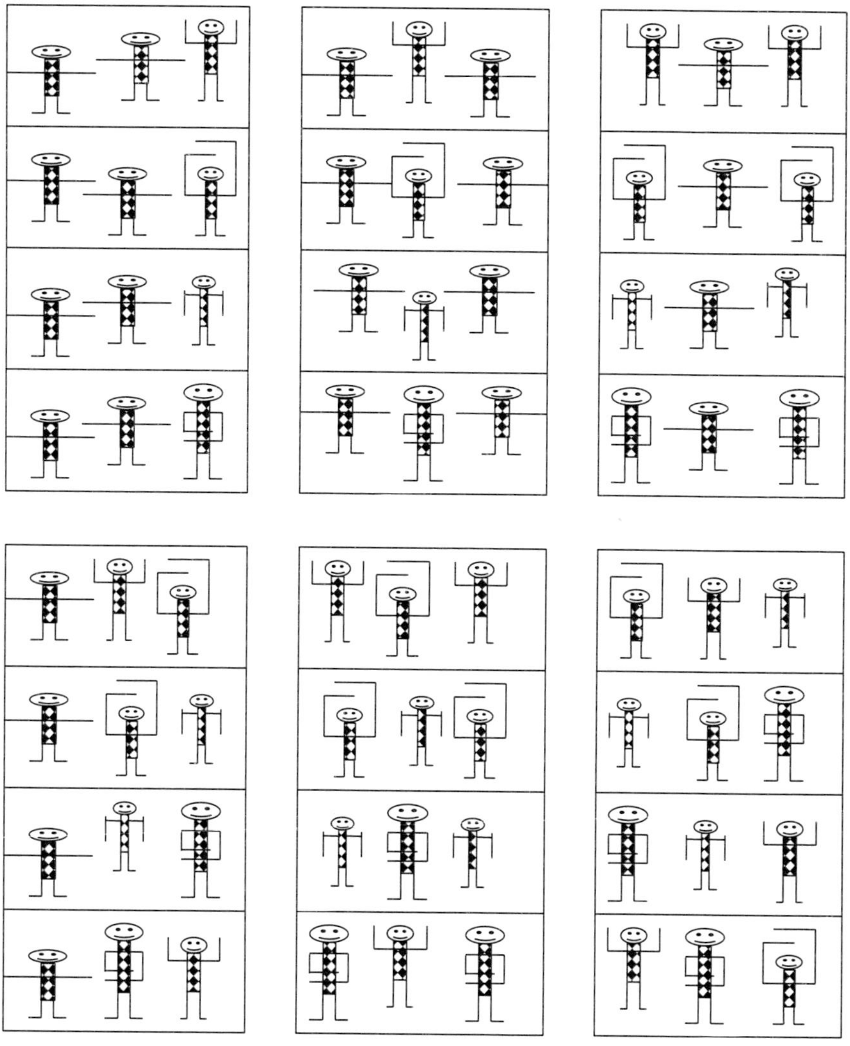

Figure 78

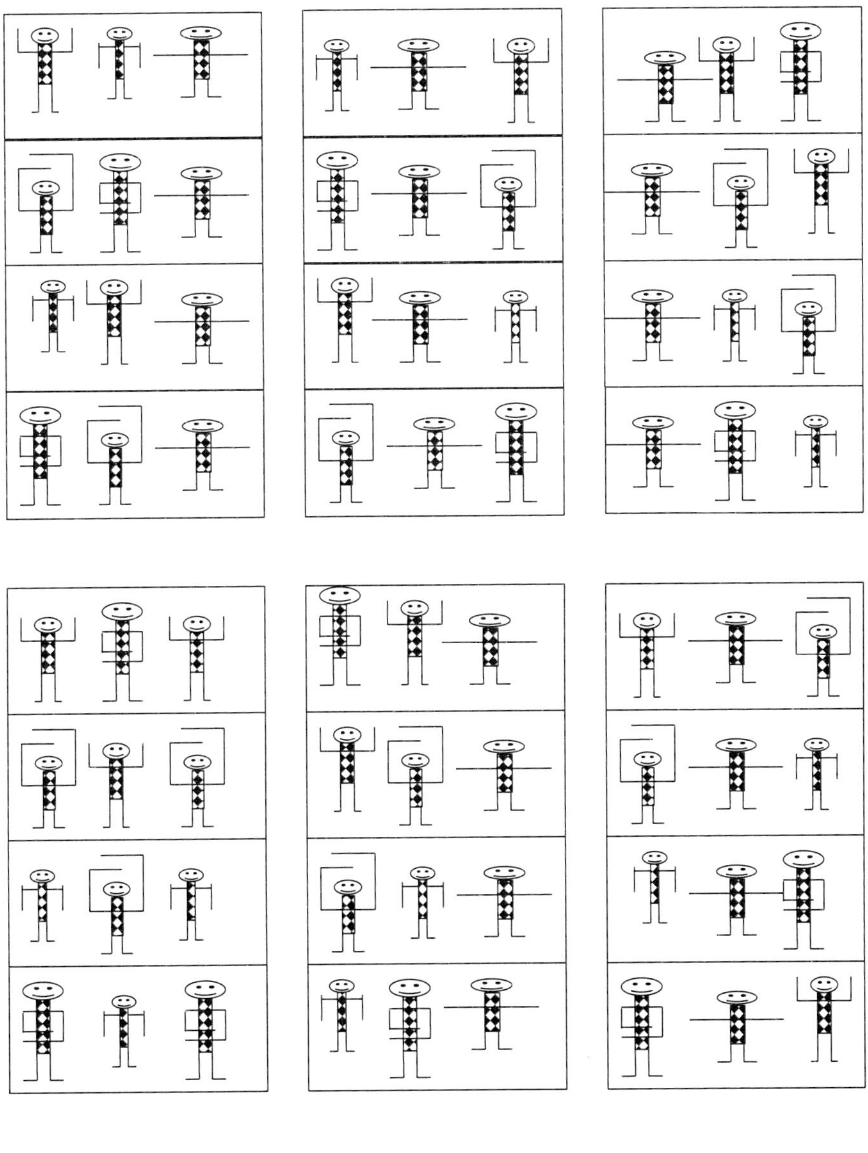

Figure 79

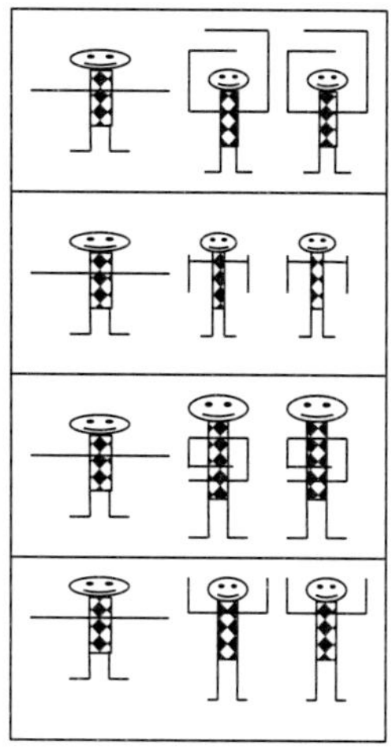

Figure 80

Every possible way of combining the positions is realised, except when all three have their arms outstretched, and each position is realised just once. The boxes have been given in a certain specific order. You might amuse yourself in trying to find out what my rule was in ordering the boxes just in the way in which I did. It is fairly obvious how I got the first and the second stick men from the previous box, my rule for getting the third stick man is a little harder to spot.

2. Instructions for Making the Boxes

Cut out the 31 rectangles described above, or draw them to your own satisfaction, adding your artistic zeal for making the stick men a little less stick-like! You could then glue them on cardboard and fold them along the horizontal lines, and then stick the lower end of the fourth picture to the upper end of the first picture with Scotch tape. You can in this way turn them into rectangular boxes, but of course there will be a hole, in the shape of a square, at each side, on your left and on your right respectively.

As you roll one of these boxes away from you, you will notice that at any of the three positions, the position of the stick man will change according to the rule given by the arrows on the first page. In fact each stick man will be multiplied by the upper closed position at each roll. If you roll the box towards you, the stick men will get multiplied by the lower closed position. After two rolls, each stick man will assume an opposite position, upper and lower being interchanged, but open or closed being conserved. This is, of course, a multiplication by the lower open position. Leaving the box as it is corresponds to multiplying by the upper open position.

3. Capturing Boxes, Starting with Two Given Boxes

Here is an introductory game you might like to play. I shall call a face of the box with three stick men on it a picture.

Pick any two boxes. You will have picked eight pictures. Add any two of these eight pictures. By this is meant that you add the first stick man of one picture to the first stick man of the other picture, to get the first stick man of the sum picture. Do the same with the second stick men, adding second to second, and the same with the third stick men, adding third to third. You may or may not already have the picture you have obtained as your sum. If you do not yet have it, find it. The box with your sum picture on it has been captured. Place it by your first two boxes.

You can now use pictures on the captured box, or you can go on using pictures from your starting boxes to get more sum pictures. Every time you get a sum that you do not as yet have in your collection, find the box with this picture on and capture it.

After a while you will find that you keep getting sums that are on boxes that are already in your collection. In fact, when you already have six boxes in your collection, you will never get any more sums outside your collection. The hunt is over; you cannot capture any more boxes! If you have more than six boxes, you have made a mistake in adding!

Let us call such a set of six boxes a family.

Once you have collected a family, have a good look at all the 24 pictures in your family and see if they have something in common, which non-members do not have! If you find such a property of the boxes in your family, you could give your family a name, using the property. You must make sure not only that all family members have the property in question, but that no other boxes outside the family have it!

For example, you might have started with the first stick man on each of your first two boxes always having his arms outstretched. Then there is no way in which you could ever capture any other box but one on which the first stick man has his arms outstretched. This is because zero + zero = zero.

Or you might have started with two boxes on both of which the first stick man is in the same position as the second stick man. In this case all your captured boxes will also have first and second stick men in the same positions. You might call the first family the 'left zero' family. The second one could be called the 'left equals middle' family.

There are 31 families that can be constructed in the way described. Each family can have a suitable name, which distinguishes its members from non-members. Try to find as many families as you can and give them suitable names. What happens when you add a picture to itself?

Or if you add two pictures that are on the same box?

You could play the game competitively, if you prefer. Two players play in this case, each player taking 2 boxes at random from a pile of 31 boxes.

Each player tries to capture all the boxes he or she can, as fast as possible.

Speed might be important, as there will always be a box which belongs to both families, namely to the family being constructed by one player as well as to the family being constructed by the other player. It is, of course, possible that one of the players has picked a box that belongs to the family that the opposing player is trying to construct. The worst case scenario will be if all four boxes picked belonged to the same family! Then each player will be madly looking for the two missing boxes, which each player will be requiring in trying to complete the family, which of course is a lost cause, because the best that either player can do is to have four of the six boxes in the family.

The game can become faster and more challenging when all the 31 names have been found and learned. Any two boxes already determine the family, and it is a lot easier to find members if one knows the name by which members can be identified!

4. Possible Names for the Families

Here are some possible names for families, with brief explanations as to why a name might be suitable.

(1) Left zero. The middle and the right-hand men are the only ones that can be in non-zero positions. The left-hand man does not move his arms as you roll the box.
(2) Middle zero. Similar reasons.
(3) Right zero. Similar reasons.
(4) Middle equals under of right. The middle man and the right-hand man on the same picture are opposites, and so each man will be in the position of the under of the other.
(5) Middle equals back of right. The middle man and the one at the back of the right-hand man are in the same positions.
(6) Middle equals front of right. The middle man and the one at the front of the right-hand man are in the same positions.
(7) Middle equals right. The middle man and the right-hand man are in the same positions.
(8) Left equals under of right. The left-hand man and the right-hand man on the same picture are opposites, so each man will be in the position of the under of the other.
(9) Left equals back of right. The left-hand man and the one at the back of the right-hand man are in the same positions.
(10) Left equals front of right. The left-hand man and the one at the front of the right-hand man are in the same positions.
(11) Left equals right. The left-hand man and the right-hand man are in the same positions.
(12) Left equals under of middle. The left-hand man and the middle man on the same picture are opposites, so each man will be in the position of the under of the other.
(13) Left equals back of middle. The left-hand man and the one at the back of the middle man are in the same positions.
(14) Left equals front of middle. The left-hand man and the one in the front of the middle man are in the same positions.
(15) Left equals middle. The left-hand man and the middle man are in the same positions.
(16) Total zero. If you add all three men in the same picture, you get the sum zero.
(17) Middle plus left equals back of right.
(18) Middle plus left equals front of right.
(19) Middle plus left equals right.
(20) Left plus right equals back of middle.
(21) Left plus right equals front of middle.
(22) Left plus right equals middle.
(23) Right plus middle equals back of left.
(24) Right plus middle equals front of left.
(25) Right plus middle equals left.
(26) Left plus front of middle equals front of right.

(27) Left plus back of middle equals back of right.
(28) Left plus front of middle equals right.
(29) Left plus under of middle equals back of right.
(30) Left plus back of middle equals right.
(31) Left plus under of middle equals front of right.

I thought that by the time you got through sixteen names, it should be easy enough to figure out the reasons for the rest of the names! You might be able to find better, more suitable names for your families. Just go ahead and see if you can improve on my list!

To practise the names you could play the following game:

Pick any two names and find the box that has both those names, in other words the box that belongs to both the families identified by the names you have chosen. You could then find the names of the other four families to which this box belongs. Alternatively, pick two boxes and find the name of the one family to which both boxes belong. You could then find all other members of that family. Every family has six boxes in it and every box belongs to six different families.

5. The 5 × 5 Square Problem

You will find the following problem quite a challenge!

Choose 25 of your 31 boxes and place them in 5 rows or 5 boxes, but in such a way that all boxes in the same row should belong to one and the same family, but also so that all boxes in the same column should belong to one and the same family.

To make life a little easier, I suggest that you make sure that the 6 boxes that you are not going to use should form a family. For example, you might like to remove all the boxes in the 'total zero' family, and use the remaining 25 boxes for solving the problem.

I shall use a more concise notation for the stick men for the purpose of explaining problems in which many of the boxes are involved. I shall use 0 for the horizontally outstretched arm position. The open positions will be denoted by the letter A, the closed positions by the letter B. I shall use upper case for upper positions and lower case for lower positions.

One picture will be enough for identifying a box, as we now know what the other pictures on the box are going to be by thinking of rolling the box backwards and forwards. For example, the 'total zero' family could be identified by means of these six pictures:

[AAb], [bAA], [AbA], [OAa], [AaO], [aOA]

Table 38 shows how the remaining 25 boxes can then be arranged.

Table 38

AAA	OOA	AAB	AAO	Aaa
Abb	ABb	OAO	AOb	Aab
AOO	ABA	bBA	Aba	OAb
OAA	ABa	AOB	AaA	AbO
Aaa	ABO	AbB	OAB	AOA

You should now be able to work out the names of the families that have their members in the rows and the names of those that have their members in the columns. The box AAb belongs to all the row families and the box bAA belongs to all the column families.

Where are the families, in the above figure, to which each of the other four boxes belongs?

6. Sequences of Boxes

It is quite easy to make up rules for deciding which box should come after which in a sequence of boxes. For example, as shown in Table 39, here is a rule that brings in all the boxes, involving just one picture of each box.

Table 39

Picture you are looking at	Next picture
Position of man in the middle	Position of man on the left
Position of man on the right	Position of man in the middle
Sum of left, middle and right	Position of man on the right

You can start with any of the 124 pictures. Let us start with the box on which all three positions are equal, say with the picture AAA. The sequence will begin like this:

AAA, AAb, AbO, bOa, OaB, aBA, BAB, ABO

and so on and will return to AAA when it has been through all the boxes!

You can make rules that take in more than one picture of each box involved. Table 40 shows a rule that will take in two pictures from each box.

Table 40

In the picture you are looking at		In the next picture
Position of right-hand man	is	Position of left-hand man
Sum of middle and right	is	Position of middle man
Sum of all three men	is	Position of right-hand man

The rule will take you through all the boxes, then it will come back to the first box you chose and start going through all the opposite positions of the first round.

You might be wondering whether it is possible to find a rule that will produce a sequence of all the pictures. This is indeed possible. Here is one such rule:

under of left + under of middle + front of right = left next
under of middle + front of right = middle next
under of left + front of right = right next

The order in which the boxes are listed in the instructions for making them also follows one of these rules. The rules, suggested by Roy Davies of Leicester, are shown in Table 41.

Table 41

Position of middle	Is	Position of left next
Position of right	Is	Position of middle next
Front of left plus middle	Is	Position of right next

Here is the sequence as it goes through all the pictures:

AAA, AAb, Abb, bb0, b0a, 0aA, aAa, Aaa, aaA, aAB, ABa, Baa, aab, abB, bBA, BAb, Ab0, b00, 00A, 0A0, A0A, 0AB, ABA, BAa, Aa0, a0A, 0Ab, AbA, bA0, A0B, 0BB
BBB, BBA, BAA, AA0, A0b, 0bB, bBb, Bbb, bbB, bBa, Bab, abb, bbA, bAa, AaB, aBA, BA0, A00, 00B, 0B0, B0B, 0Ba, BaB, aBb, Bb0, b0B, 0BA, BAB, AB0, B0a, 0aa
aaa, aaB, aBB, BB0, B0A, 0Aa, AaA, aAA, AAa, Aab, abA, bAA, AAB, ABb, Bba, baB, aB0, B00, 00a, 0a0, a0a, 0ab, aba, baA, aA0, A0a, 0aB, aBa, Ba0, a0b, 0bb
bbb, bba, baa, aa0, a0B, 0Bb, BbB, bBB, BBb, BbA, bAB, ABB, BBa, BaA, aAb, Aba, ba0, a00, 00b, 0b0, b0b, 0bA, bAb, AbB, bB0, B0b, 0ba, bab, ab0, b0A, 0AA

and then back to AAA.

There are some amusing facts that we can gather from the above cycle. Start with any two consecutive pictures, for example the pictures AaA and aAA, and find the family to which both of them belong. In this case the family is 'left equals under of middle', since the left-hand man and the middle man are opposites. You will find the remaining four members of the family by following this rule:

Using the letter Y for 'being a member of the family' and the letter N for 'not being a member of the family', the boxes will appear in the sequence:

Y Y N Y N N N N Y N N N Y N N N N N Y N N N N N N N N N N
N N

repeating four times.

So all the 31 families will appear one after the other in the same sequence of members and non-members. You might check and see whether the same thing is true of the sequence of pictures produced by the first rule I suggested. If it is, you might wonder if it would be the case with all rules that take in all the pictures! Or if not, what other rules might there be according to which the families appear to be organised within a full sequence taking in all the pictures!

I will leave you to ponder over these things.

You might ask a mathematician friend, if you have one, to help you solve this or any other problems that might occur to you as a result of playing with the 31 boxes.

Chapter XVI

FUNCTIONS IN A 3 × 3 UNIVERSE

1. Construct Your Mini-Universe

Place nine blocks in three rows of three as shown (see Figure 83). To each block associate two numbers, which will be the first and the second numbers. To calculate the first number of a block use the following codes:

thin = 0, thick = 1, small = 0, big = 1

and add them.

To calculate the second use these codes: circle = 0, square = 1, triangle = 2

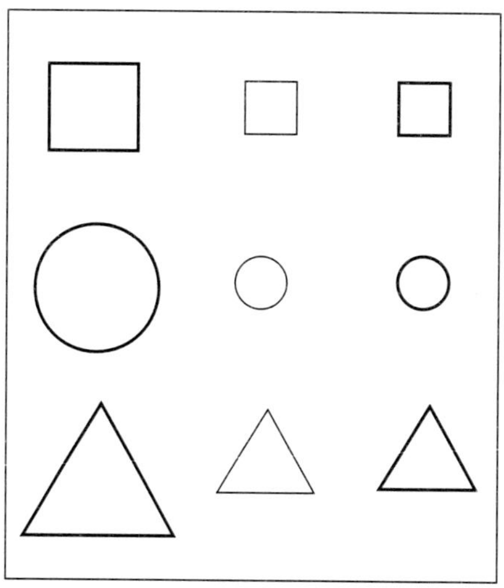

Figure 81

Each block will thus have two numbers. For example, the top left corner will be referred to as 21. Table 42 shows a map of the pairs of numbers we shall use.

Table 42

21	01	11
20	00	10
22	02	12

The numbers never reach beyond 2. For this reason, if in any calculation using any rules you obtain a number greater than 2, you must simply subtract 3 from it.

2. How to Get Around

Figure 82 shows a rule for going from one block of the diagram to another block (just choose a number pair, then go from left to right to get the next number pair).

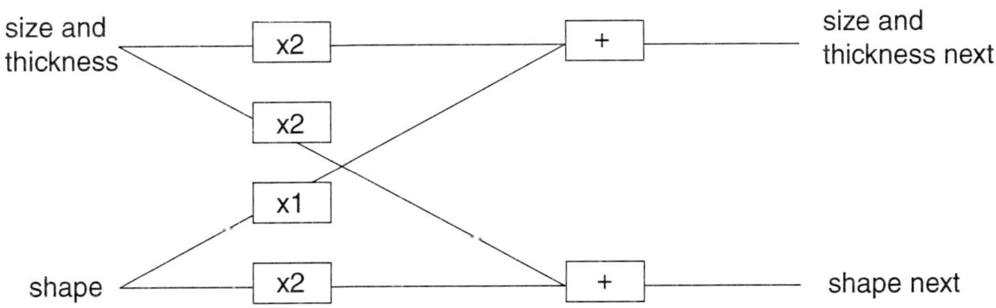

Figure 82

Try not to forget to subtract 3 every time you obtain a number greater than 2!

Try to work out how you will get from one block in the block diagram to another, following the above rule. You will find that you will have taken in eight blocks, all except the small thin circle, and you will come back to your starting point.

Here is your path through your diagram, each step from block to block shown with an arrow:

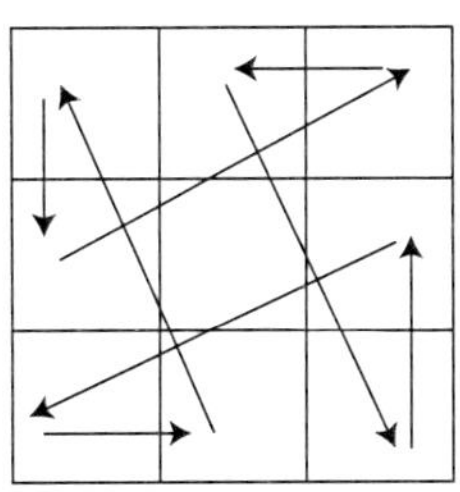

Now invent your own rule. All you have to do is to fill in the blanks after the multiplication signs in the figure below. The multipliers can be 0, or 1 or 2.

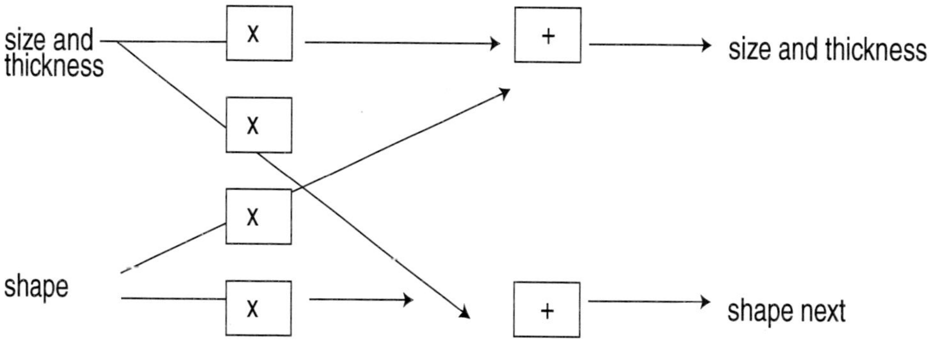

Most rules will not take you all over the block diagram. Some will make you go back and forth between two spots, some will give you cycles of three, some cycles of four and some cycles of six. Some rules might tell you to stay put at certain blocks! If you find a rule that takes you to the small thin circle, then the same rule will never allow you to leave!

The table below shows you how to find two ways of getting round your diagram. Try to find out what the rules are that allow you to go from block to block according to these ways.

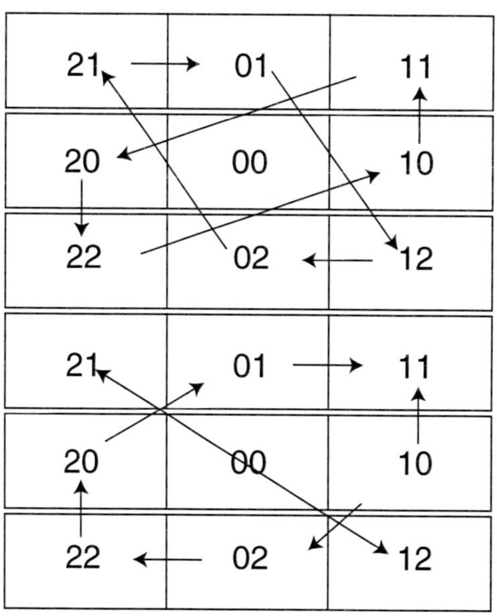

Figure 83

Here are two rules placed end to end:

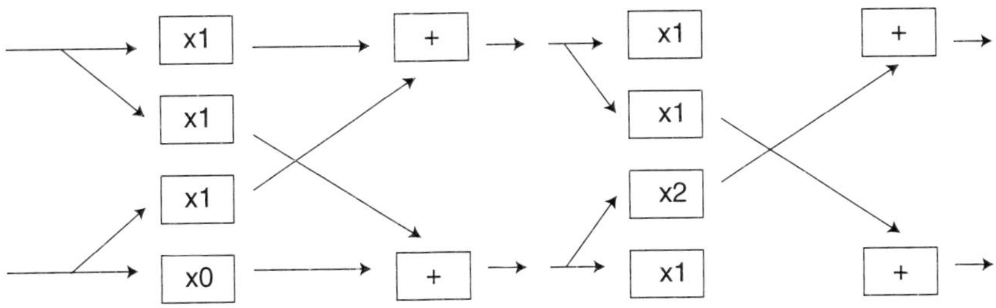

which can be replaced by the rule:

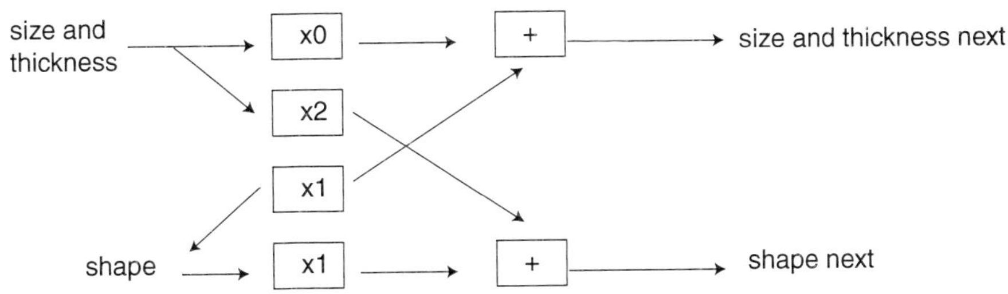

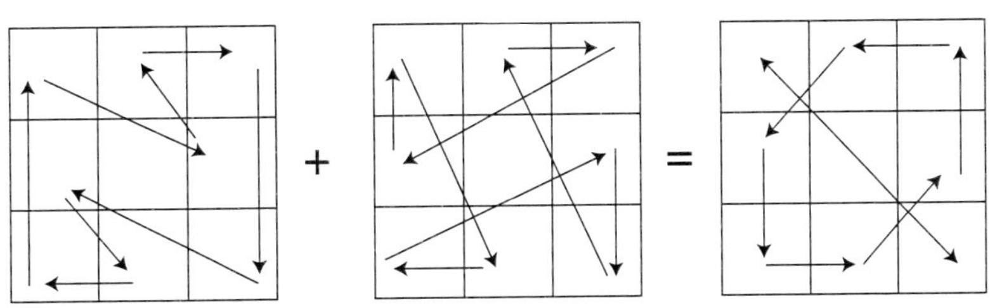

Figure 84

Try to find the rule that you would have to use if you followed the first rule and then the second at each step. Would it be the same rule as doing the second rule followed by the first rule at each step?

There are 48 possible rules for getting round the block diagram which never take you to the small thin circle. Any two of these rules, when applied one after the other, will give a resultant rule for getting round the block diagram. Such a resultant rule will always be one of the 48 rules. So the system is closed.

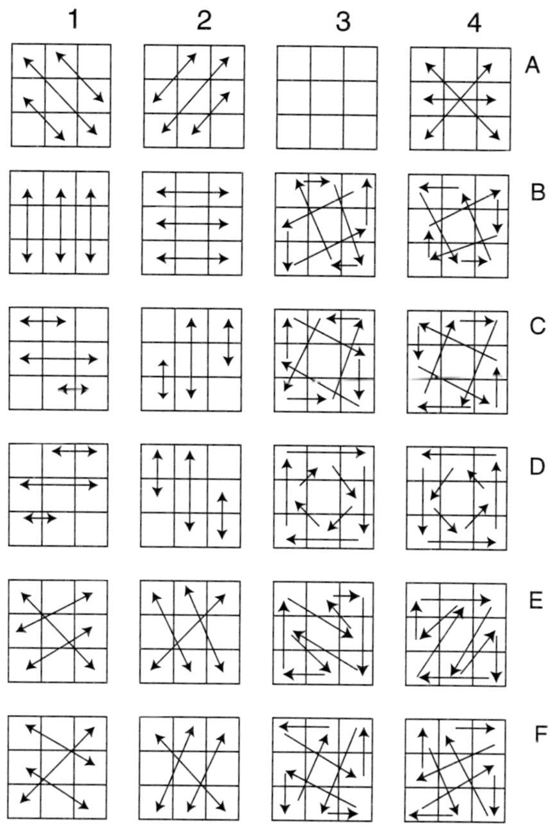
1
2
3
4
A
B
C
D
E
F

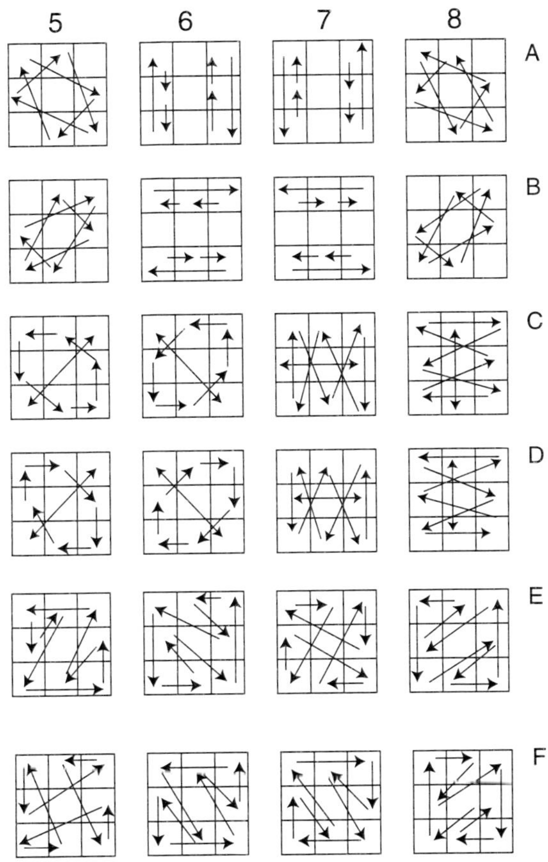

4. How to Find the Rule for a Given Way of Getting Round

Let us take one of the 48 ways of getting round listed and see how we might find its rule. Let us try the way round the block diagram given in the cell 5D (see Table 44).

Table 44

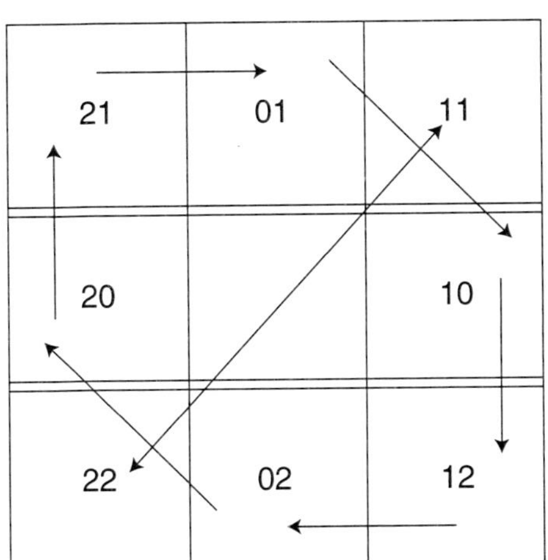

We only need two pieces of information from the walk to discover the rule. Here is one way of doing it:

Let us take the following information:

$$11 \rightarrow 22 \text{ and } 12 \rightarrow 02$$

Let us remember that the first digit tells us the size and the thickness of the block and the second one tells us its shape. We have to find out what the multiplier is by which we have to multiply the first number and the multiplier by which we have to multiply the second number so that after adding these products and subtracting any 3s, we obtain the first number of the next block. This will tell us the size and the thickness of the next block. Then we have to do the same for the second number of the next block along the way. This will tell us the shape of the next block.

Let us go for the size and thickness first. Let us denote the multiplier of the size and thickness number of the block where we are by A and the multiplier of the shape number of the block where we are by B. From the two pieces of information we are given, we have:

$$(1 \times A) + (1 \times B) = 2 \text{ and } (1 \times A) + (2 \times B) = 0$$

Whatever B might be, $3 \times B$ must always become 0, as after all the 3s have been subtracted there will be nothing left since $3 \times B$ is a multiple of 3. Making use of this useful fact, we can add the above equations and get:

$$(2 \times A) + (3 \times B) = (2 \times A) + 0 = (2 \times A) = 2$$

and so we must have $A = 1$. Since we know that $A + B = 2$, we must have also $B = 1$. This means that simply adding the size and thickness number of the block where we are to the shape number of the block where we are, we must get the size and thickness number of the next block on the way. It is easy to check that this is so (do not forget that you must take away any 3s at the end).

To find the rule for the shape of the next block, we can call the multipliers C and D respectively. The equations to solve will now be:

$$(1 \times C) + (1 \times D) = 2 \text{ and } (1 \times C) + (2 \times D) = 2$$

Adding the equations again, this time we get:

$$(2 \times C) + (3 \times D) = (2 \times C) = (2 + 2) = 1$$

and so $C = 2$ (since $2 \times 2 = 1$, as we must remove a 3) but since $C + D = 2$, we must have $D = 0$.

This means that 2 multiplied by the size and thickness number of the block where we are will give us the shape number of the block where we are going. It is easy to check that this is so. So we have:

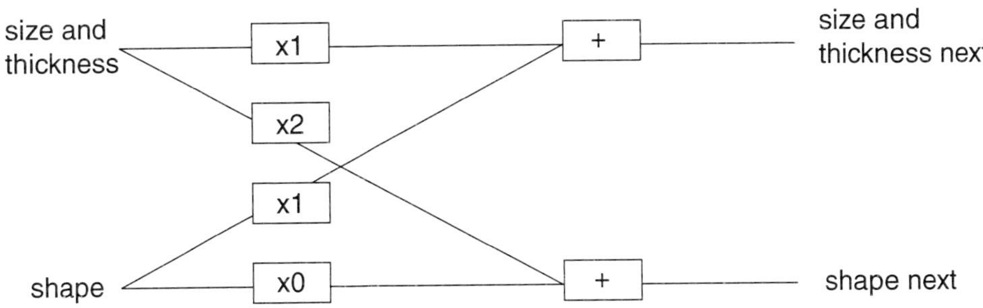

5. Algebraic Language

Let us review the language we have been using. The multipliers we put in the four boxes of our rule diagrams will tell us what rule we have to follow for getting round our arrangement of blocks.

To find the size and thickness number of the block we are going to we have to multiply the size and thickness number of the block where we are by one of the multipliers; we also have to multiply the shape number of the block where we are by another multiplier. Then we have to add these products, and finally subtract as many 3s as we can from the sum. This will give us the size and thickness number of the next block.

To find the shape number of the block we are going to, we also have to multiply the present size and thickness number as well as the present shape number by the given multipliers and add these products. Again we must subtract as many 3s as possible from this sum. This will give us the shape number of the next block.

Let the multipliers be called A, C, B and D, taken vertically from the highest to the lowest in the rule diagram. The letters A and B will be names for the multipliers for getting size and thickness next. The letters C and D will be names for the multipliers for getting shape next. Let the size and thickness number of the block where we are be called T, and the shape number of the block where we are be called S.

Let the size and thickness number of the next block be called t, and the shape number of the next block be called s. We can express what we are doing as follows:

$$AT + BS = t$$

$$CT + DS = s$$

You might now find it an interesting challenge to find a way of expressing a product rule, whenever you multiply a rule by another by carrying out one of them and then the other one.

It would be good to call the size and thickness numbers and the shape numbers of the three different blocks different names:

the size and thickness number of the block in the first spot = t_1

the size and thickness number of the block in the second spot = t_2

the size and thickness number of the block in the third spot = t_3

the shape number of the block in the first spot = s_1

the shape number of the block in the second spot = s_2

the shape number of the block in the third spot = s_3

Then we can write our equations like this:

$t_1 + s_1 = t_2$ $t_2 + 2s_2 = t_3$

$t_1 = s_2$ $t_2 + s_2 = s_3$

Putting the values of t_2 and of s_2 from the first set of equations into the second set, we have:

$$t_1 + s_1 + 2t_1 = t_3$$

and

$$t_1 + s_1 + t_1 = s_3$$

but

$$t_1 + 2t_1 = 3t_1 = 0$$

and so we can write our equations as:

$$s_1 = t_3$$
$$2t_1 + s_1 = s_3$$

which is in fact the rule that takes us from our first block to our third block.

If we change the order in which we apply the rules, we shall sometimes get a different product rule. Taking one rule first and then doing the other rule we get the way given in cell 6C, but reversing the order we get the way given in cell 7D. You might like to check if I am correct in saying this.

In textbooks on mathematics you would see the rules written respectively as:

$$\begin{matrix}1\,1\\ \\1\,0\end{matrix} \quad \times \quad \begin{matrix}1\,2\\ \\1\,1\end{matrix} \quad = \quad \begin{matrix}0\,1\\ \\2\,1\end{matrix}$$

the first row giving the multipliers for the size and thickness next, the second row the multipliers for the shape next, the multipliers being given in the order size and thickness first, shape second.

Using such substitutions, we can always calculate the product rule out of any given two rules. The product rule is the single rule that will always take us to the same block as the successive application of our two chosen rules. But do not forget to say which rule you are using first!

If you are curious to know more about such rules, take a book out of the library on linear algebra and read the chapter on matrices. You can then find the mechanical rule for combining two rules, if you do not wish to use your brain.

You can now draw the maps of your chosen ways of getting round the block diagram, as long as it is one of the 48 ways given in the list! For example, you could try to find the rules for these ways of getting around:

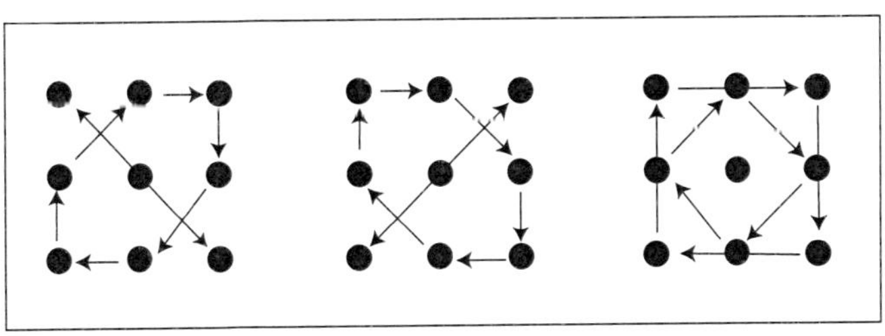

6. Finding Some Old Friends Among the 48 Functions

You might have been wondering whether any functions with which you might already be familiar are included in the set of 48 functions whose maps I have listed. The answer is that you will find a surprising number of them among the 48 functions in our list! In what follows I shall try to unravel in what way we can recognise our old friends, now that they have been dressed up as functions acting on a 3×3 universe!

Let me take a look at our various cycles first. If we look over the list, we will find cycles that have the following lengths:

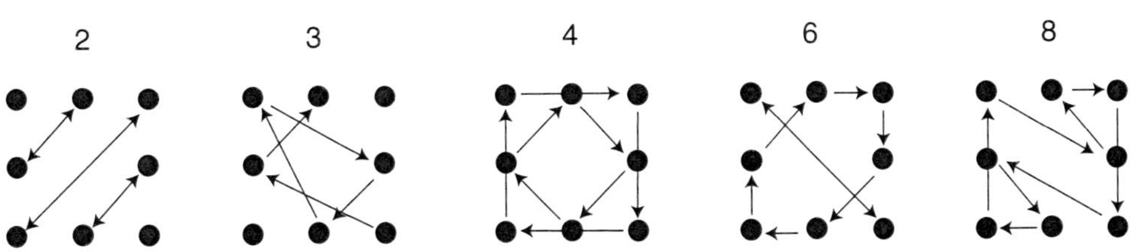

Here are some generators of cycles, followed by all members of the cycles:

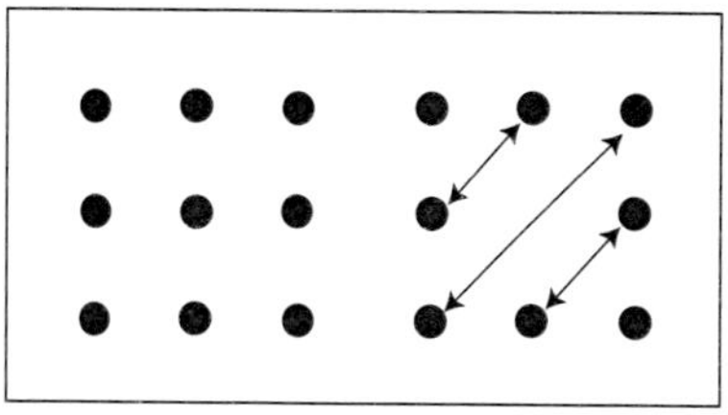

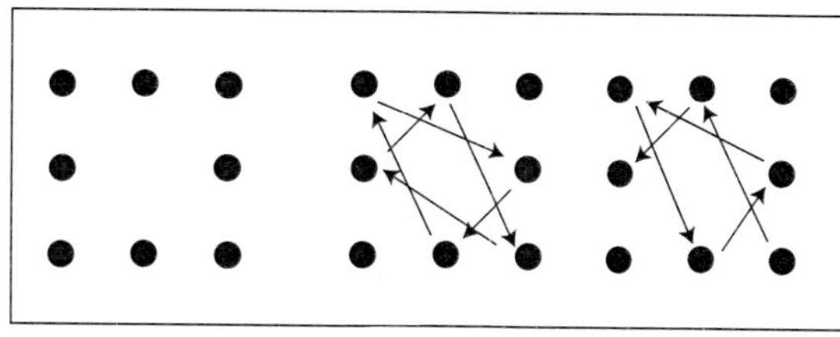

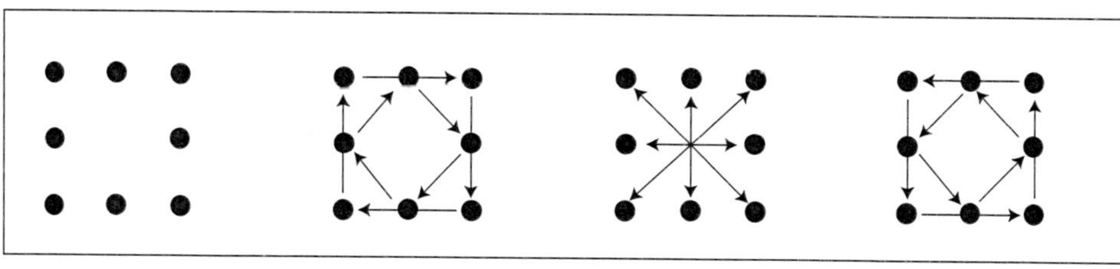

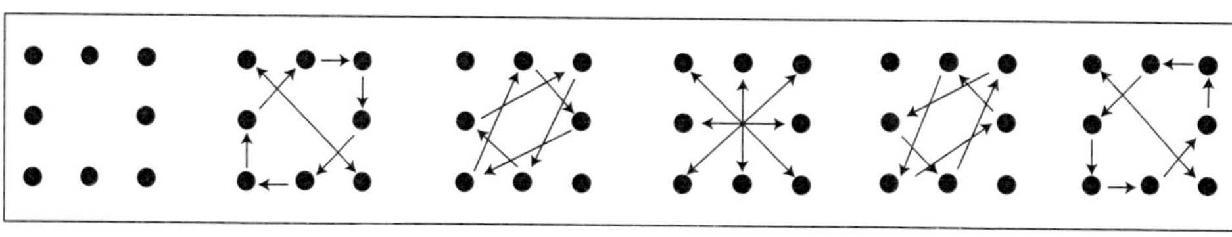

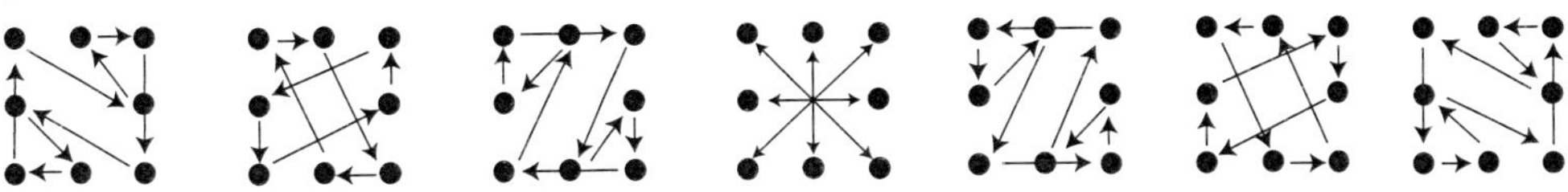

1 2 3 4 5 6 7

The first chart is the generator in each example. By applying the generator twice we obtain the diagram given under the number 2, applying it three times we obtain the one given under the number 3, and so on.

We have already found one of our sets of four functions: they form the cycle of four. But here is another set of four:

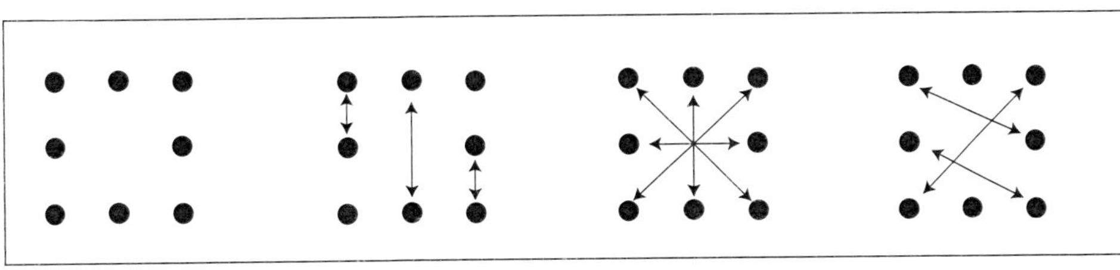

0 1 2 3

If you carry out any two out of 1, 2 and 3, it will have the same effect as carrying out the third one. Is this familiar?

Now let us try to find 6 functions among our 48 which will work in the same way as the rotations and the reflections of the equilateral triangle.

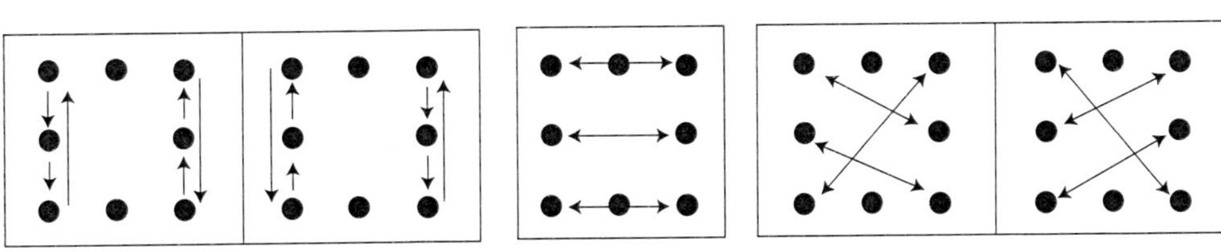

$\begin{matrix}1&0\\1&1\end{matrix}$ $\begin{matrix}1&0\\2&1\end{matrix}$ $\begin{matrix}2&0\\0&1\end{matrix}$ $\begin{matrix}2&0\\1&1\end{matrix}$ $\begin{matrix}2&0\\2&1\end{matrix}$

where our 5 non-neutral functions are also given in the conventional way in the form of 2 × 2 matrices, as mathematicians call them.

Below you will find the functions we can call the square functions (less the neutral); they relate to each other in just the same way as do the rotations and reflections of the square.

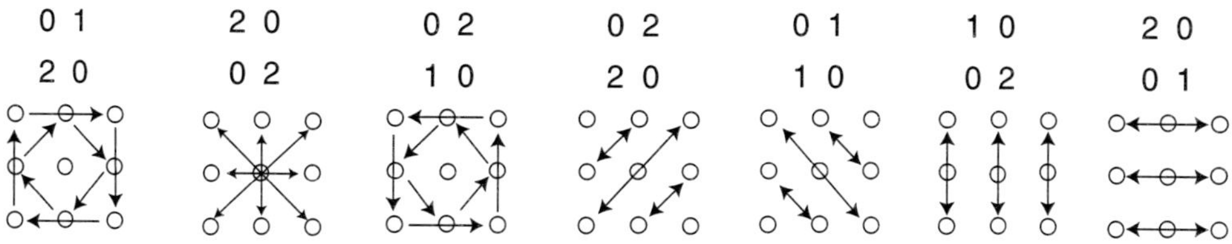

The first and the third functions correspond to quarter turns of the square, while the second function corresponds to a half turn. The other four functions correspond to the reflections about the axes of symmetry of the square.

Next we can move to finding the 12 functions corresponding to the rotations and reflections of the regular hexagon. Here is one way in which we can make 12 of our 48 functions correspond to the 12 isometries of the regular hexagon:

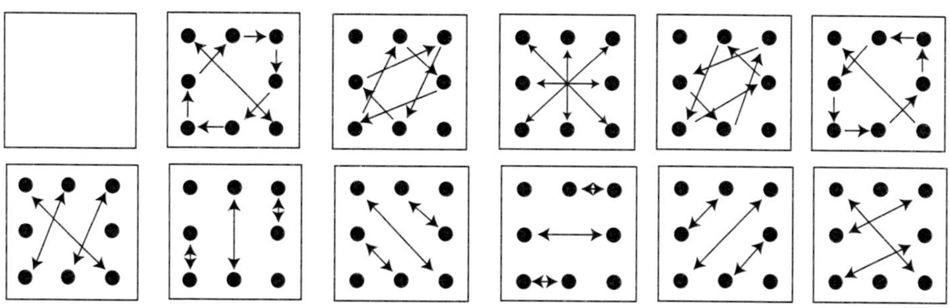

The charts in the first row represent the rotations, while those in the second row represent the reflections of the regular hexagon.

The conventional expressions of these functions are shown in Table 45.

Table 45

10 01	01 21	21 20	20 02	02 12	12 10
12 02	10 12	01 10	21 01	02 20	20 21

There is one more interesting set of 8 functions which I have kept until last. These are the eight functions that make up the so-called quaternion group. These include all the functions in our set of 48 that are four cycles. Here they are:

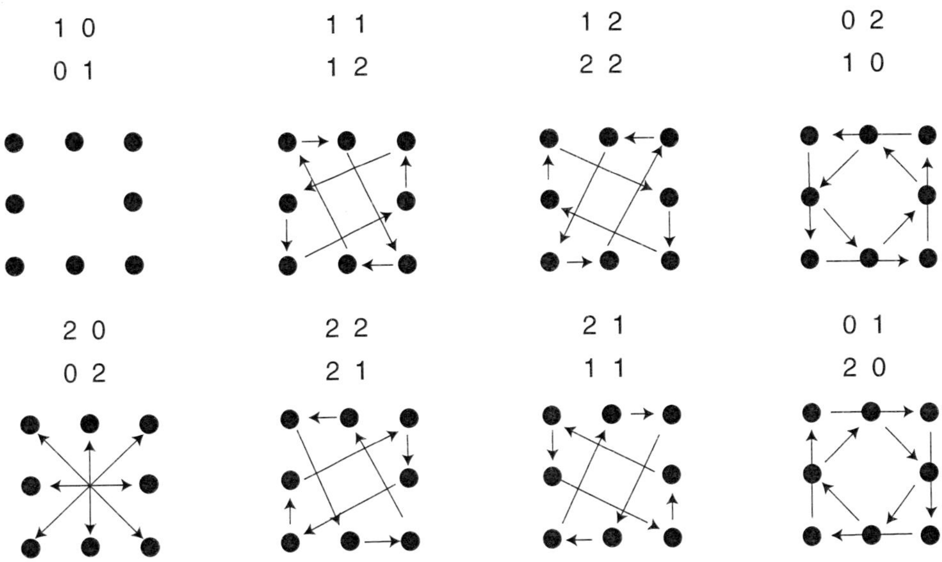

Let us call the functions in the first row 1, i, j and k and the functions in the second row –1, –i, –j, –k respectively.

Try to check that the following combination rules are true:

$$ij = k,\ jk = i,\ ki = j,\ ji = -k,\ kj = -i,\ ik = -j,\ ii = jj = kk = -1$$

and that multiplying by –1 will simply change pluses into minuses and vice versa. The functions 1, i, j and k are the units of the quaternion algebra!

Chapter XVII

THE TWENTY-SIX PICTURES GAME

1. Instructions How to Make our Toys

Photocopy these double pictures onto stiff paper or card, then cut them out and fold each one at the dividing line between the two pictures. You can then stand them up so that each picture appears the right way up. You will have thirteen such double pictures; these will be our playthings for the following activities.

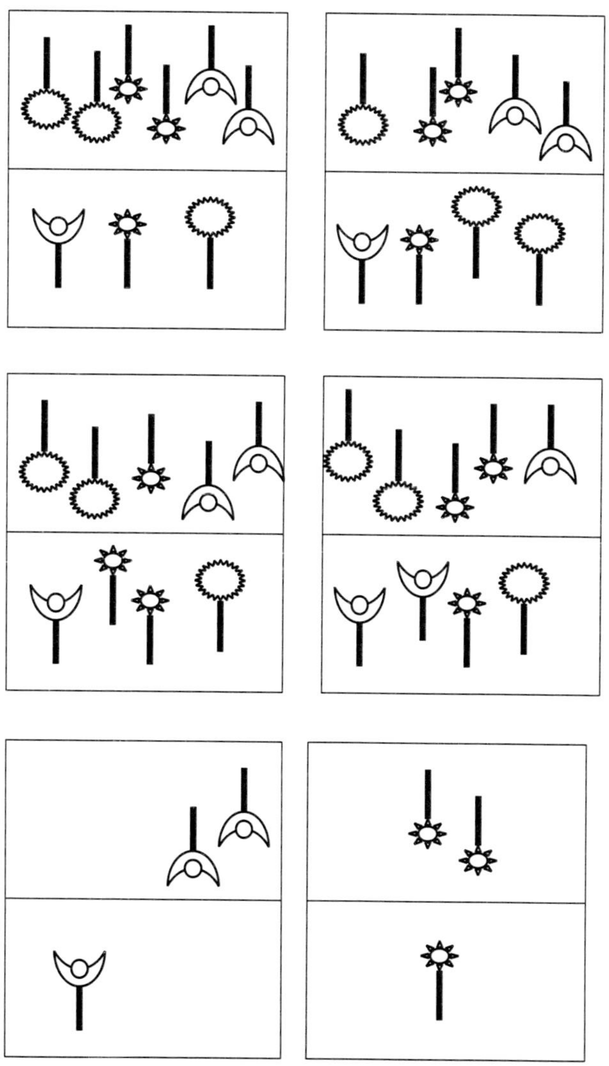

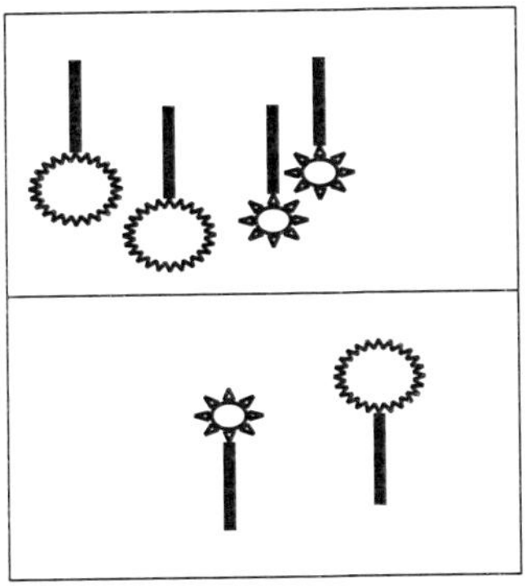
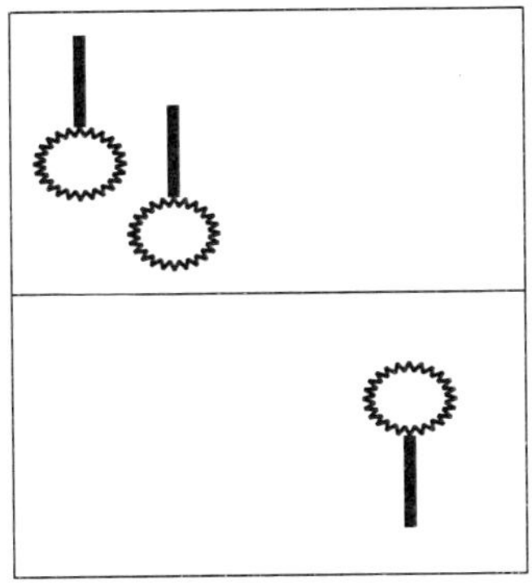
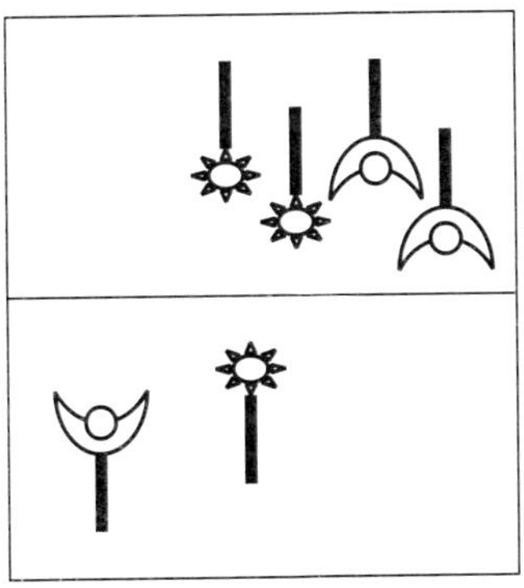
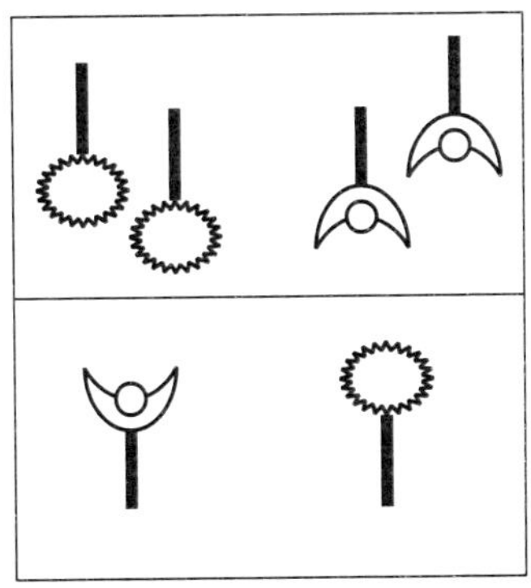
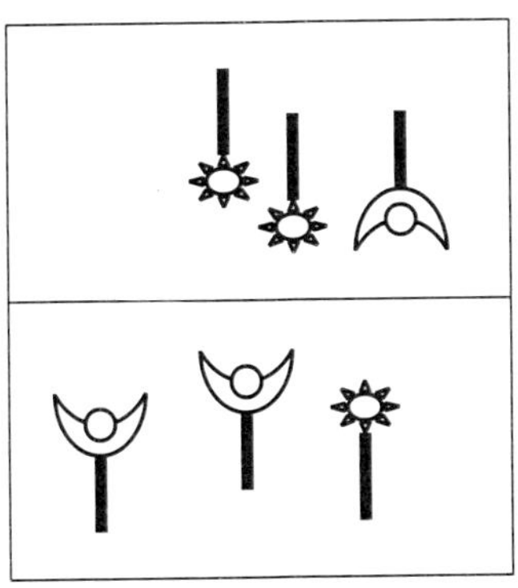
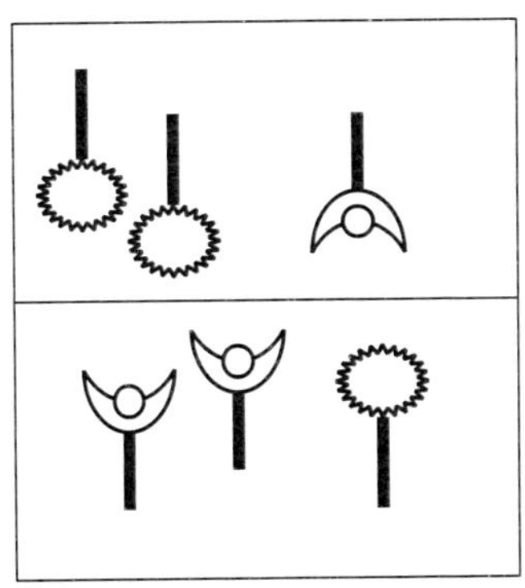
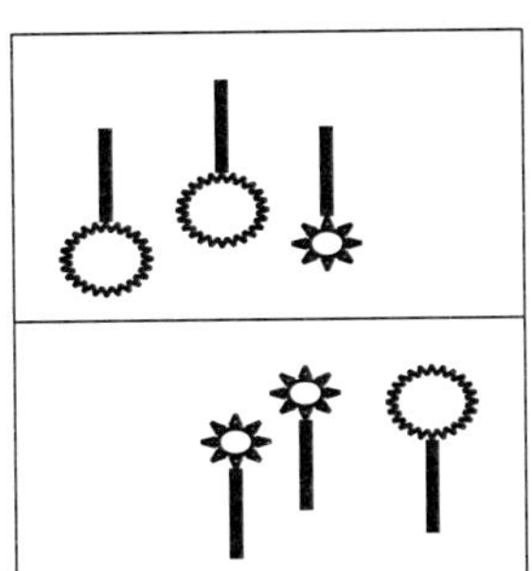

You will have noticed, if you have looked through the pictures, that there are thirteen pictures the right way up and another thirteen upside down. This is so that if you cut out each double picture, you can fold along the line separating them, so by standing them up both pictures will appear the right way up.

I shall call the big round flowers sunflowers, and the small round ones daisies. I think the third kind look quite like tulips! You will also no doubt have noticed that in each set of two pictures, the numbers 2 and 1 are interchanged. I shall regard a total of three of any kind of flower as 0. So the total number of flowers on any set of two pictures that have been drawn together is then 0. In fact the only numbers that will be allowed in our games are: 0, 1 and 2.

2. Addition and Multiplication

Addition of our three numbers is done almost like the normal addition, except that

$$2 + 2 = 1$$

because out of the 4 we would normally get as our sum, three numbers are counted as 0, so instead of 4, which is not one of our three numbers, we have the number 1.

Multiplication is also almost the same as in normal multiplication, with the exception of 2×2, and again we have, for the same reasons as before,

$$2 \times 2 = 1$$

I shall describe each picture simply by stating how many of each kind of flower can be found on the picture. I shall count the tulips first, then the daisies and then the sunflowers. For example, [1, 1, 2] will mean that there is one tulip, one daisy and two sunflowers in the picture; [2, 0, 2] will mean that there are two tulips, no daisies and two sunflowers in the picture.

3. Adding Pictures

We can add a picture to another picture by simply adding tulips to tulips, daisies to daisies and sunflowers to sunflowers.

Let me call each double picture, each toy, a prism. You could actually make the toys look more like prisms by making triangular prisms out of cardboard and sticking the double pictures on two of the faces of the cardboard prism. You will be using the third face of the prism as its base on which it can stand.

If you have cut out and suitably bent your toys into prisms so you have all the pictures the right way up, you can play the following game:

(1) Pick any two of your prisms.
(2) Choose one picture on one of the prisms and another picture on the other.

(3) Add these chosen pictures.
(4) Find the prism which has this picture on it.
(5) We say that you have captured this third prism.
(6) Place it with your two chosen prisms, thus making an enlarged set.
(7) Now choose any two pictures from prisms in your enlarged set.
(8) Add these pictures and find the prism with it on and so capture it.
(9) Go on like this capturing prisms by adding more and more pictures.
(10) When you have four prisms in your set, any further sums will be in this set.
(11) We can call such a set of four prisms a family of prisms.

Try to see in what way the eight pictures on your four prisms are alike. The way they are alike must be such that none of the pictures outside your set is like them. Such a property can then be used as the name of the family.

For example, if neither of your first prisms has any sunflowers, then there is no way you can capture any prisms with sunflowers on it. So such a family might be called the 'no sunflowers' family.

You may have chosen both prisms so that in all four pictures the number of daisies is equal to the number of tulips. In adding your pictures, you will only get other sum pictures with the same number of daisies as tulips. You could call such a family the 'daisies same as tulips' family.

There are actually thirteen families you can find by starting with two given prisms. Try to find as many as possible of these families. Give each family a suitable name. Do not forget that the name must apply to all members of the family, but must not apply to any pictures outside the family.

Here is an example. Suppose you choose the prisms:

1 2 1		1 2 0
2 1 2		2 1 0

Let us find the sum 121 + 120 = 211, so we capture the prism:

2 1 1
1 2 2

But then what about 1 2 1 + 2 1 0 = 0 0 1? We have captured:

0 0 1
0 0 2

If we try any more adding of pictures we already have, we shall again get one of the pictures we already have. So we have a family. What could be the name of this family?

Except for the last prism captured, we always have a different number of tulips and daisies. If you look in the next section, you will find such a family listed at the seventh place. We could make the strange convention that 0 is equal to 0 as well as different from 0, in which case all four prisms have the property of having a different number of tulips from the number of daisies.

If you don't like such an absurd convention (and you would be in good company), you will need to change some names. For example, we could say that in this family, if we add the number of tulips to the number of daisies, we always get 0. Maybe it is not quite so absurd to call the numbers 3 and 0 equal as to call zero the same as itself as well as different from itself! You can make your choice, or maybe you can find an even more suitable name!

4. Suitable Names for the Thirteen Families

Here are some suggested names for the families that you can find. You are welcome to improve on the names:

(1) no tulips
(2) no daisies
(3) no sunflowers
(4) tulips same as daisies
(5) daisies same as sunflowers
(6) tulips same as sunflowers
(7) tulips different from daisies
(8) daisies different from sunflowers
(9) tulips different from sunflowers
(10) tulips plus daisies equals sunflowers
(11) daisies plus sunflowers equals tulips
(12) tulips plus sunflowers equals daisies
(13) total number of flowers zero

5. Make a 3 × 3 Pattern with Nine Prisms

Here is a fun problem you might like to solve. I give a solution later on, but don't look if you want the fun of finding out for yourself.

(i) Choose nine of your prisms.
(ii) Place them in three rows of three prisms.
(iii) Make sure, however, that prisms in the same row all belong to one family.
(iv) Also make sure that prisms in the same column all belong to one family.

(v) For good measure, make it so for the diagonals as well!

One useful hint would be to choose your nine prisms in such a way that the four prisms not being used should all belong to one family. If you are not careful with this, you will have difficulties!

What is the fourth member of each row family? Of each column family? Of the diagonal families? Are the missing members all in the family outside your pattern?

With the outside family you will have used nine families in total: three row families, three column families, two diagonal families and the outside family. There are four more families! Where are they in your pattern?

Given any two families, do they always have a common member? Given two prisms, is there just one family to which they both belong? To how many families does each prism belong? If you had a picture with no flowers on it at all, would it belong to any of the families? If so, to which ones would it belong?

Here is a solution to the 3 × 3 pattern problem:

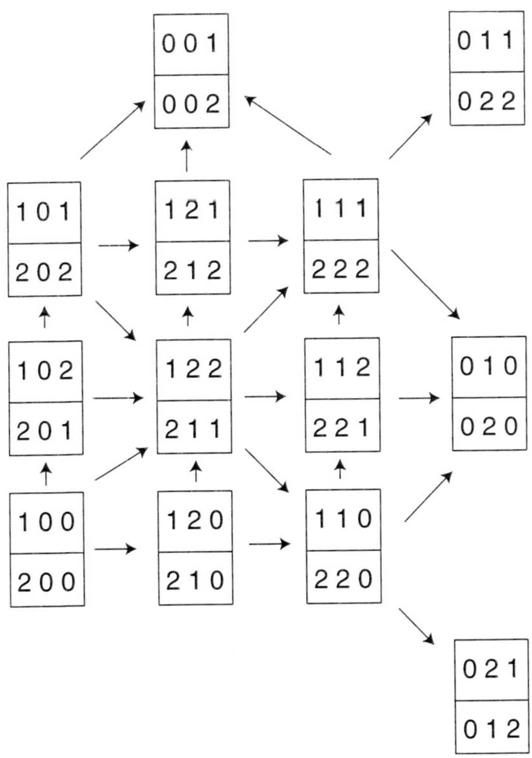

The four prisms left out of the pattern form the family with no tulips. These prisms are:

(i) only sunflowers
(ii) daisies same as sunflowers

(iii) only daisies
(iv) daisies and sunflowers equals 0

The first of these is a member of each of the three column families, the second one completes one of the diagonal families, the third one is a member of each of the three row-families and the fourth one completes the other diagonal family.

The four missing families are along the sub-diagonals with the prism in the opposite corner in each case. The names of these families are:

(a) tulips plus sunflowers equals daisies
(b) daisies plus sunflowers equals 0
(c) tulips plus daisies equals sunflowers
(d) total number of flowers equals 0

6. Bundles

Every prism belongs to four families, and every family consists of four prisms. Any two families have one prism in common, and any two prisms belong to just one family. Four families all having a prism in common form a bundle.

A bundle game could be played like this:

(i) All thirteen family names and all thirteen prisms are placed randomly on the table.
(ii) One player picks up three names at random.
(iii) Players have to determine whether these names belong to a bundle or not.
(iv) The first player who is sure will shout either 'bundle!' or 'no bundle!'
(v) A mistaken declaration can be challenged by the other players.

Naturally the game can also be played by a player picking three prisms at random, the other players having to decide whether the three prisms do or do not all belong to one and the same family. The shouts then would be 'family!' or else 'no family!'

An interesting bundle problem is the following:

(i) Place all the prisms in a tidy row in front of you.
(ii) By each prism place a name, not necessarily a name belonging to that prism.
(iii) Change the positions of the prisms and/or of the names so that any three prisms belonging to one and the same family should have by them three names that are a part of a bundle.

Don't look, but here is one possible interesting solution:

001	010	100	110	101	011
002	020	200	220	202	022
S = 0	D = 0	T = 0	T + D = 0	T + S = 0	D + S = 0

120	102	012	112	121	211
210	201	021	221	212	122
T = D	T = S	D = S	T + D = S	T + S = D	D + S = T

7. Making Up Rules for Sequences of Pictures

It is quite easy to make up simple rules through the use of which the prism coming after any given prism can be determined. Table 46 shows one suggestion:

Table 46

The picture you are looking at		The next picture
Tulips plus daisies	will give	tulips
Tulips plus sunflowers	will give	daisies
Tulips	will give	sunflowers

Do not forget our addition rules! From any number you get by adding which is greater than 2, you have to subtract 3 to get the sum!

This is how the sequence looks:

(001), (010), (100), (111), (221), (102), (101), (121), (021), (210), (022), (220), (122)

and back to (001) again.

You will notice that only one side of each prism occurs in the sequence. If you started with (002) instead of with (001), you would get all the other sides but none of the above.

It is interesting to look for those pictures here that belong to one and the same family. Look at any two pictures with just one picture between them. After the second one of these, jump over two pictures and you will get two more straight after the jump that belong to the family to which your first two belong. In other words in any run of seven pictures you will have this pattern:

Yes No Yes No No Yes Yes

where Yes means 'belongs to a given family' and No means 'does not belong to this given family'. For example for the 'No daisies family' you will need to look at the first seven pictures, the first, the third, the sixth and the seventh pictures belonging to this family, while no other picture in the sequence does so. You might also have noticed that if you add any three pictures in a row, you always get the fourth picture.

Try to make up your own sequences, inventing different rules and see how the families are placed within your sequences.

Table 47 shows another set of rules. These rules generate a sequence taking in all the pictures. The word 'behind' will mean look on the other side of the prism and see how many flowers there are there of the kind being referred to.

Table 47

The picture you are looking at		The next picture
Tulips behind plus daisies	will give	tulips next
Daisies plus sunflowers behind	will give	daisies next
Tulips plus daisies	will give	sunflowers next

You will find that the rules take you through the thirteen prisms, then the sequence returns to the back of the first prism with which you started and goes through the back of all the prisms, in the same order. You might also find that the rule [Yes, No, Yes, No, No, Yes, Yes] also works in this case.

Try to find other rules that take you through all the pictures and see in what order the members of any one family come up as you go through the sequence.

Chapter XVIII

CYCLES OF SEVEN AND SIX USING ARMS AND MUSIC

[This section is tough going!]

1. Task 1

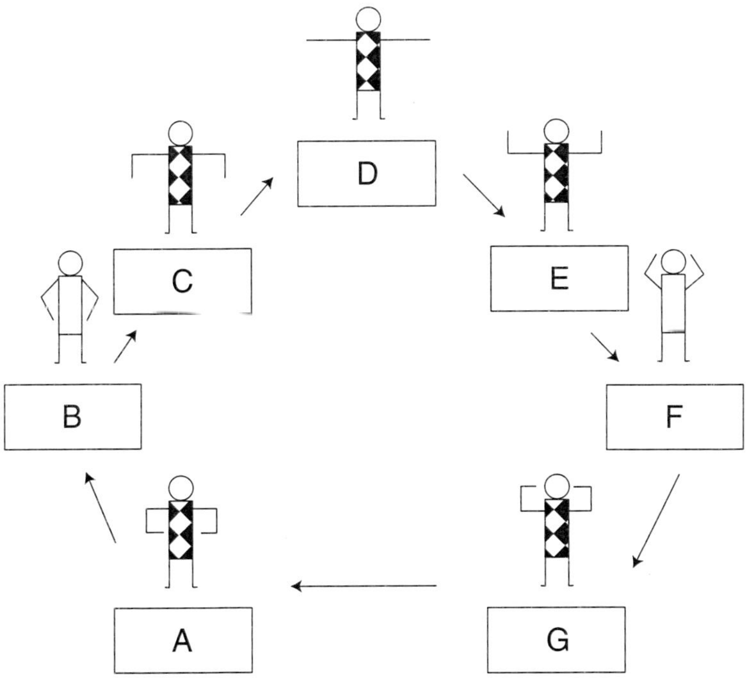

Try to practise going through the above arm positions in the order of the arrows. For positions A and G have your lower arms right on top of one another. In all the positions except in position D the lower and upper arms should be at right angles to each other. So if you start with position A, with your arms closed in the low position, you keep opening up more and more, keeping your right angle though, until you reach position D when you stretch your arms out on both sides. Then you make your right angles again and move upwards, until you are in the upper closed position at position G.

If you have someone who can play music on the recorder or on the piano, this person can play the notes from A to G, while the rest of you go through the motions as described, also humming the notes at the same time.

2. Task 2

Now try to learn the following sequence of positions:

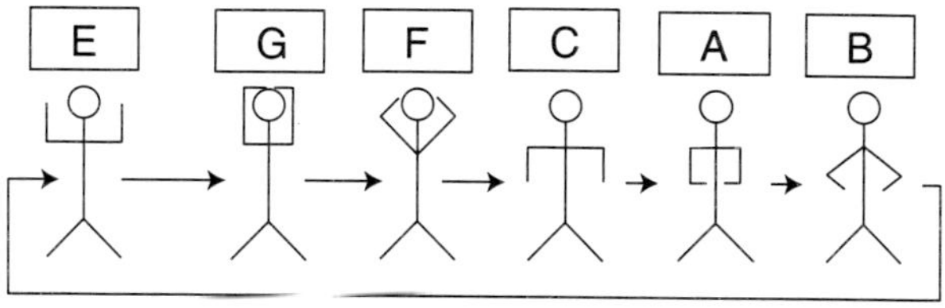

and hum the tune shown above the arm positions.

I shall call this the G cycle. The G cycle backwards I shall call the B cycle. Humming the same note all the time, I shall call the E cycle. Any note followed by a succession of Ds, I shall call a D cycle. A C cycle alternates between upper and lower positions of the same shape. So we have either E and C, or G and A, or F and B alternating. The F cycle is either E to F to A to E or G to C to B to G. The A cycle is either E to A to F to E or G to B to C to G.

Try to learn these sequences of positions, making sure you hum the corresponding note at each position! You may need to be helped out with someone playing the notes on the piano or on the recorder. You could pick one of the sequences as a first subject in a composition and write some variations to it. The G cycle would make a good lament!

3. Task 3

If you know how to add and subtract and multiply, you should find this task quite easy. Associate these numbers to the positions and notes of the previous tasks:

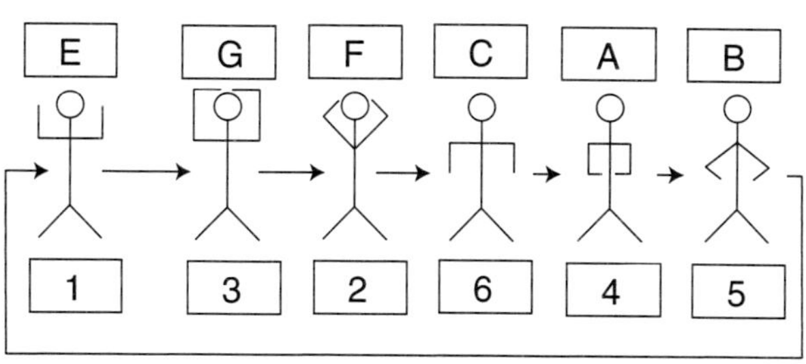

If you multiply each of the above numbers by 3, and from the product you take away 7, or 14 if you can, then you will always get the next number. So the G cycle seems to be the 3 cycle. Check this out and make sure this is so. By what number do you have to multiply one of our numbers so as to get the previous number in the sequence? Or to get the next number but one?

Try to identify each of the position-music cycles with a numerical cycle. (Don't forget to take away any multiples of seven from the products!)

4. Task 4

You can use the cycles for writing a second voice to a tune. You can use one arm for the first voice and the other arm for the second voice. The easiest one to use is the C cycle in which (E, C) and (G, A) and (F, B) are exchanged. Try *Frère Jacques*, but you should be able to jump an octave if you need to!

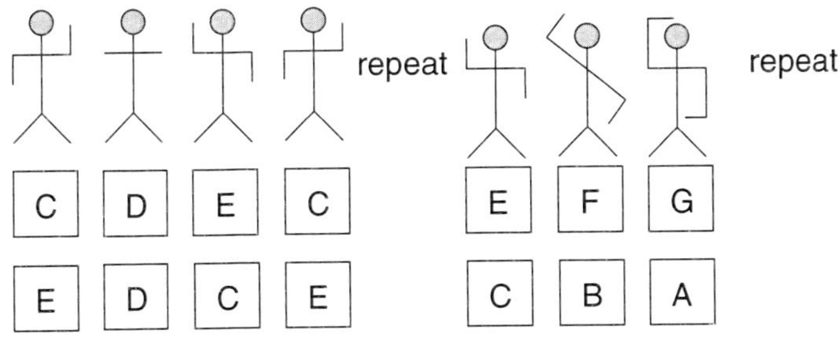

You can continue with '*Sonnez les mattines, ding, dong dong!*'

It is easier to do this with tunes that have no Fs or Bs, in other words in tunes that only have five notes. One very suitable tune is *Bonnie Banks and Braes*, beginning with:

G E C / G E C / A G E / D (first voice)
A C E / A C E / G A C / D (second voice)

You can also try to write second voices using some of the cycles suggested. You could even write a quartet, using a tune and three cycles! They make good harmonies!

5. Task 5

Let us call the horizontally outstretched arms the note D, the zero position or Z. E will be called upper open or UO, C lower open or LO. F will be called upper halfway or UH, B lower halfway or LH. G will be called upper closed or UC, A lower closed or LC.

You could now practise adding positions to each other. Table 48 shows the suggested rules for defining an adding operation (see overleaf).

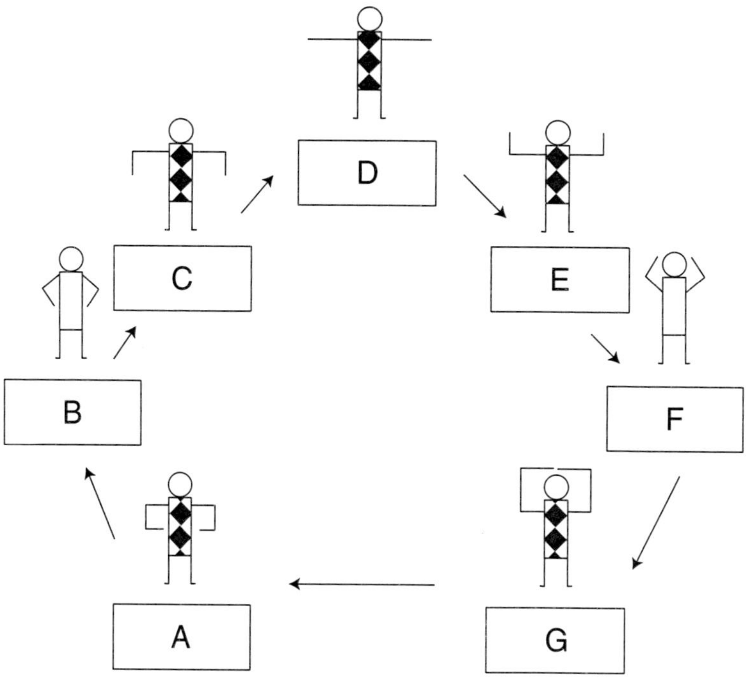

Table 48

+ Z	Position remains unchanged
+ UO	Position moves on by one arrow
+ UH	Position moves on by two arrows
+ UC	Position moves on by three arrows
+ LC	Position moves back three arrows
+ LH	Position moves back two arrows
+ LO	Position moves back one arrow

The operation (+ UO) seems to move the arms one notch further up, except when they are at the highest position, in which case the arms start again at the lowest position. The operation (+ LO) moves the arms one notch down, except from the lowest position, in which case the arms start again at the highest position.

Here is an example:

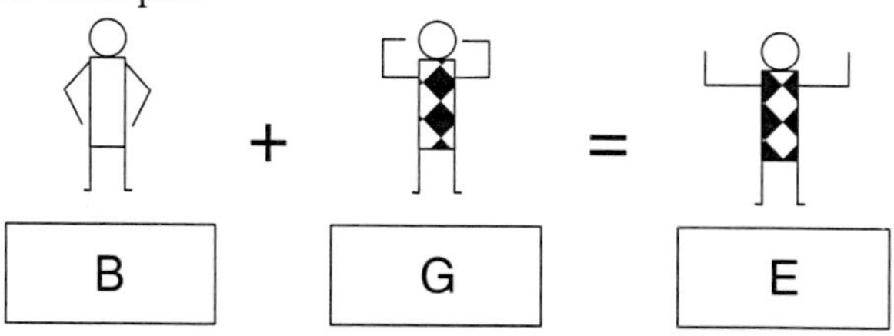

Since + G moves up B by three notches, the result is E.

6. Task 6

Here is the UC cycle, which we can use as a multiplication by UC. Multiplying by UC will move us up one notch in this cycle. For example:

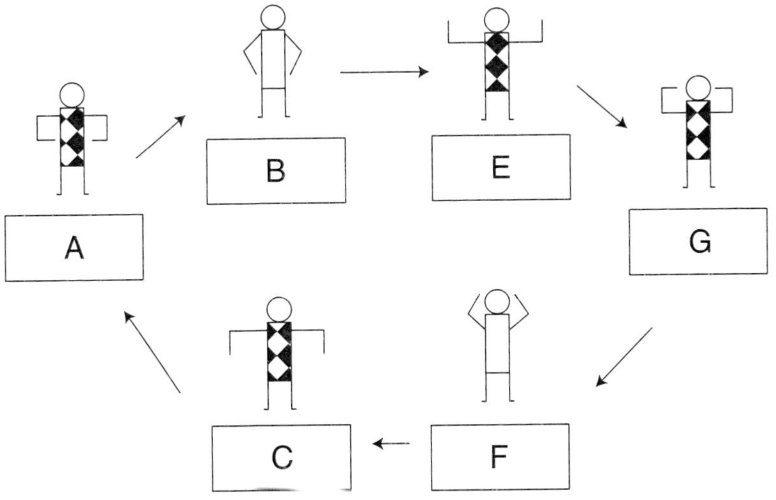

For example

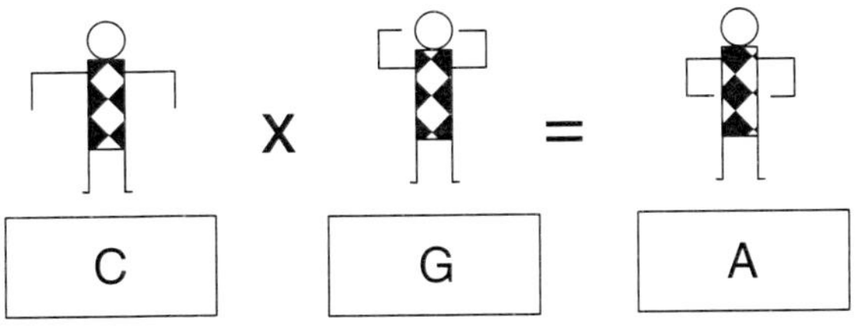

(i) Multiplying by LH will give us the opposite cycle, namely the one we obtain by going against the arrows.
(ii) Multiplying by UO will leave the multiplicand position unaltered. This is like multiplying by 1 in the normal way.
(iii) Multiplying by LO will always result in the same shape but will change upper into lower and lower into upper. It is the same as moving up three notches in the cycle.
(iv) Multiplying by UH will move us up two notches in the cycle.
(v) Multiplying by LC will move us back two notches in the cycle.

If you make the following translations:

$$Z = 0, UO = 1, UH = 2, UC = 3, LC = 4, LH = 5, LO = 6$$

you will find that by multiplying by the suggested numbers you get the same cycles as for the arm positions, assuming you take away as many sevens as you can each time from the product you get.

You can use the multiplication cycle for additions as well. Try to check whether I am right in saying the following:

(i) Adding a position to itself, you get the position two arrows on.
(ii) Adding a position to the next one, you get the position two arrows back.
(iii) Adding two positions with one position in between, you get the position in between as the sum.
(iv) Adding two positions with two other positions in between, you get the zero position.

If you can remember these tricks, you need only use one of the two cycles. Practise adding and multiplying positions, using the 'multiplying cycle'.

7. Task 7

Let us now work with the two arms moving independently of each other. Since each arm has 7 positions, we now have to deal with 49 different positions!

Let us choose any two of these positions. Then let us choose a third one. Can we, by multiplying one position by one of our multipliers, then multiplying the other position by one of our multipliers and adding the two results, reach the third position we have chosen? Here is an example. Let us start with these two positions:

If you take UO times the first one (namely the position itself) and add it to UH of the second one we get:

adding left arm to left arm and right arm to right arm.

Sometimes you might have to use Z as a multiplier. This will be the case whenever you choose your third position in such a way that both its arms are the same multiple of one or the other of the first chosen positions.

Here is a way to find the multipliers that will give us our third choice of position out of the first two:

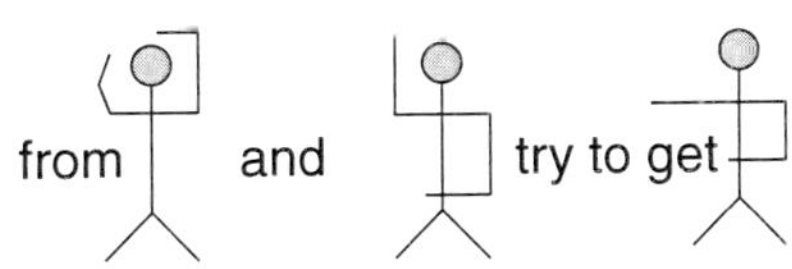

We are looking for two multipliers, one to multiply the first position, the other to multiply the second position, so that when we add the two products, we obtain precisely the third position. Let us denote these two multipliers by M and N respectively. What we have to begin with, using the symbols M and N as symbols for the unknown multipliers, is the following:

$$\mathrm{M} \star [\mathrm{UH}, \mathrm{UC}] + \mathrm{N} \star [\mathrm{UO}, \mathrm{LC}] = (\mathrm{Z}, \mathrm{LC}]$$

which has to work separately for the left arms and for the right arms, so that we have the equations to solve for M and N:

$$\mathrm{UH} \star \mathrm{M} + \mathrm{UO} \star \mathrm{N} = \mathrm{Z} \qquad (1)$$

$$\mathrm{UC} \star \mathrm{M} + \mathrm{LC} \star \mathrm{N} = \mathrm{LC} \qquad (2)$$

$$\mathrm{LC} \star \mathrm{M} + \mathrm{UH} \star \mathrm{N} = \mathrm{Z} \qquad (3)$$

We get (3) if we multiply the first equation by UH. Then adding (2) and (3), since

$$UC \star M + LC \star M = (UC + LC) \star M = Z \star M = Z$$

we have

$$(LC + UH) \star N = LC + Z = LC$$

but

$$LC + UH = LO$$

but since

$$LC + UH = LO$$

we have

$$LO \star N = LC$$

and since

$$LO \star UC = LC$$

we have

$$N = UC$$

Now, putting this newly found value for N into equation (1), we have

$$UH \star M + UO \star N = UH \star M + UO \star UC = UH \star M + UC = Z$$

Then adding LC to both sides of the last equation we have

$$UH \star M = LC$$

since

$$LC + UC = Z$$

but since

$$UH \star UH = LC$$

we see that

$$M = UH$$

So we have to UH our first position, UC our second position, and, adding the results, we should obtain our third position. You can check it, but I think it is correct!

'UH of' means move up two notches in the multiplication cycle, then 'UC of' means move up one notch in the multiplication cycle.

We have to keep in mind the multiplication table of our various multipliers. Here it is, but you should check at least some of the products. Multiplying by Z is taken in the first row and it always yields Z.

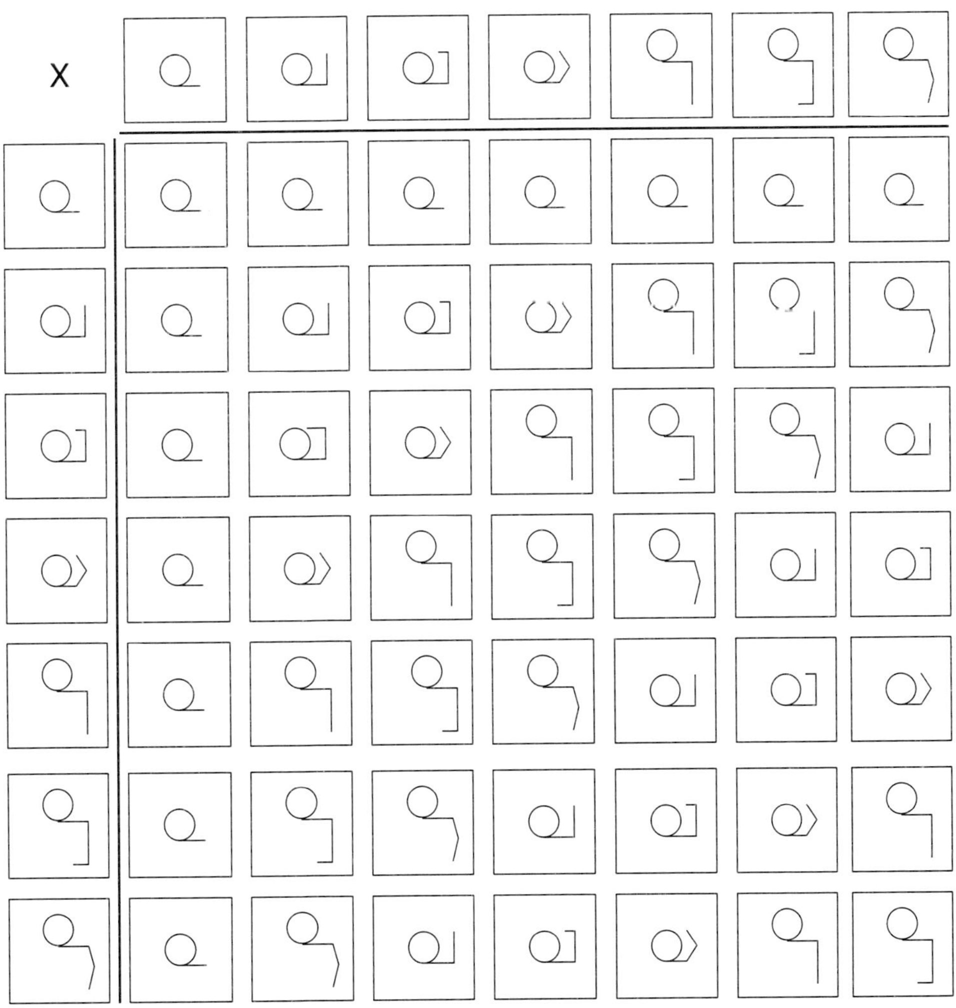

Here are some other ways of remembering what the various multipliers do to given positions:

(i) upper closed multiplier or UC: Open to closed to half then change levels and go through the same sequence of shapes.

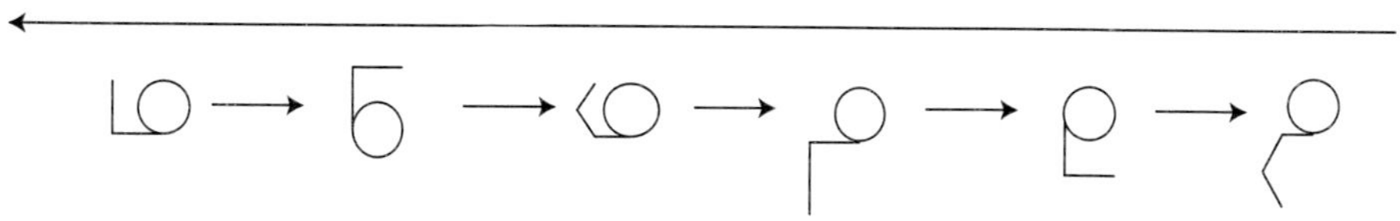

(ii) lower half multiplier or LH: half to closed to open then change levels and go through the same sequence of shapes.

UC and LH are inverses of one another.

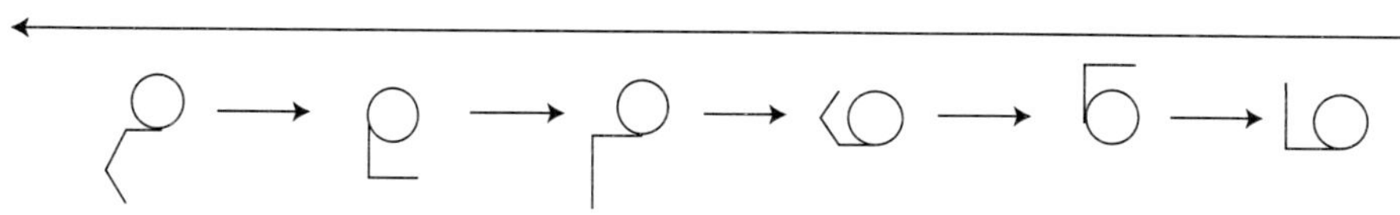

(iii) upper half multiplier or UH: open to half then change levels to closed.

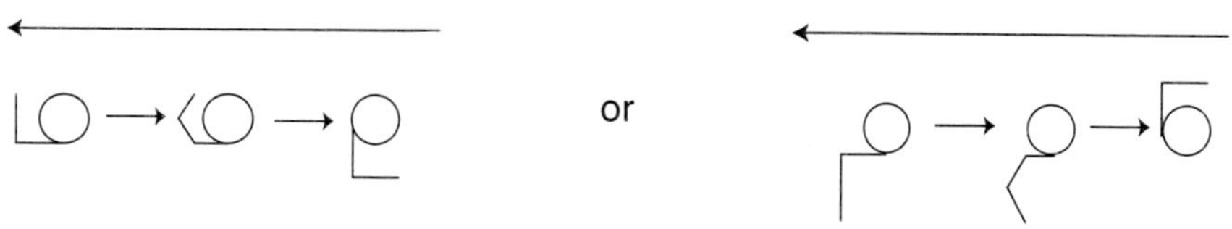

(iv) lower closed multiplier or LC: half to open then change levels to closed. UH and LC are inverses of one another.

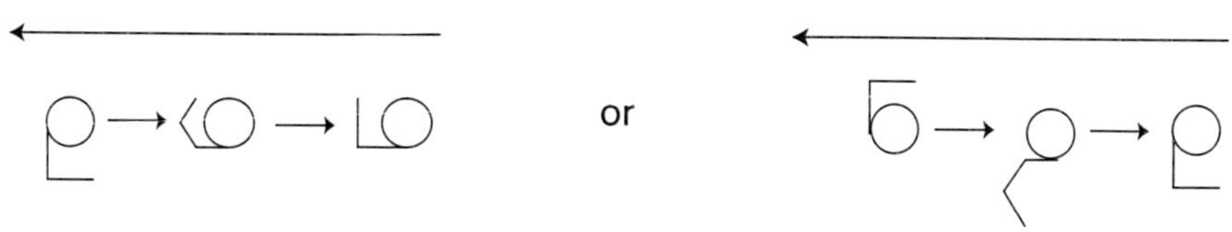

(v) Upper open multiplier or UO: no change.
UO is its own inverse.

(vi) Lower open multiplier or LO: alternate between upper and lower of the same shape. LO is its own inverse.

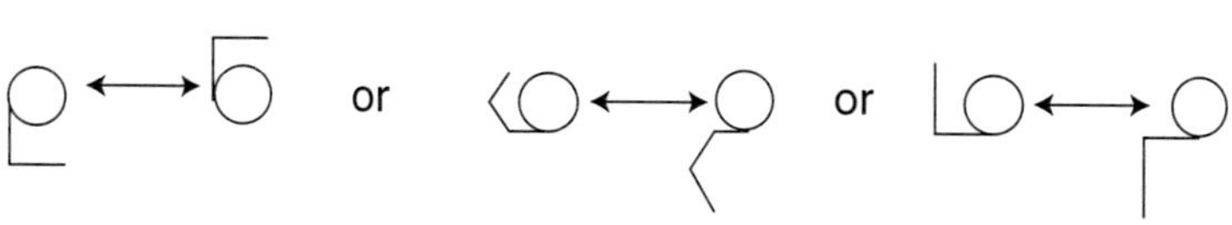

ILLUSTRATIONS TO BE CUT OUT

It would be best to photocopy any needed diagrams onto stiff paper or card

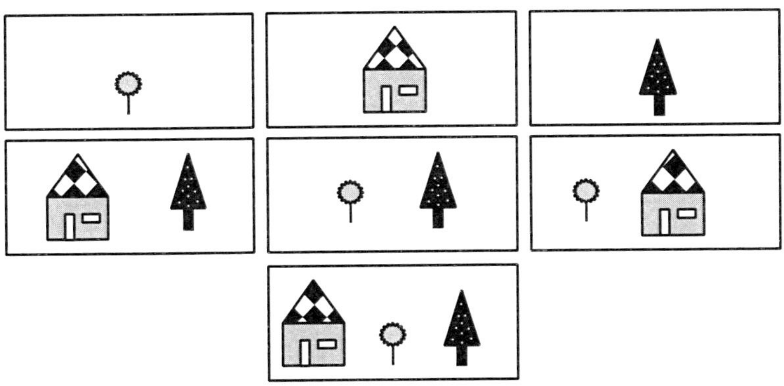

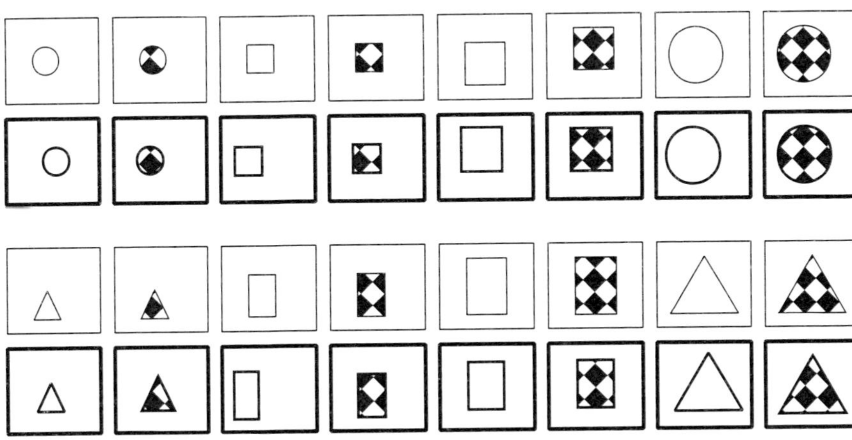

ban	bin	fan	fin	fat	fit	bat	bit
bans	bins	fans	fins	fats	fits	bats	bits

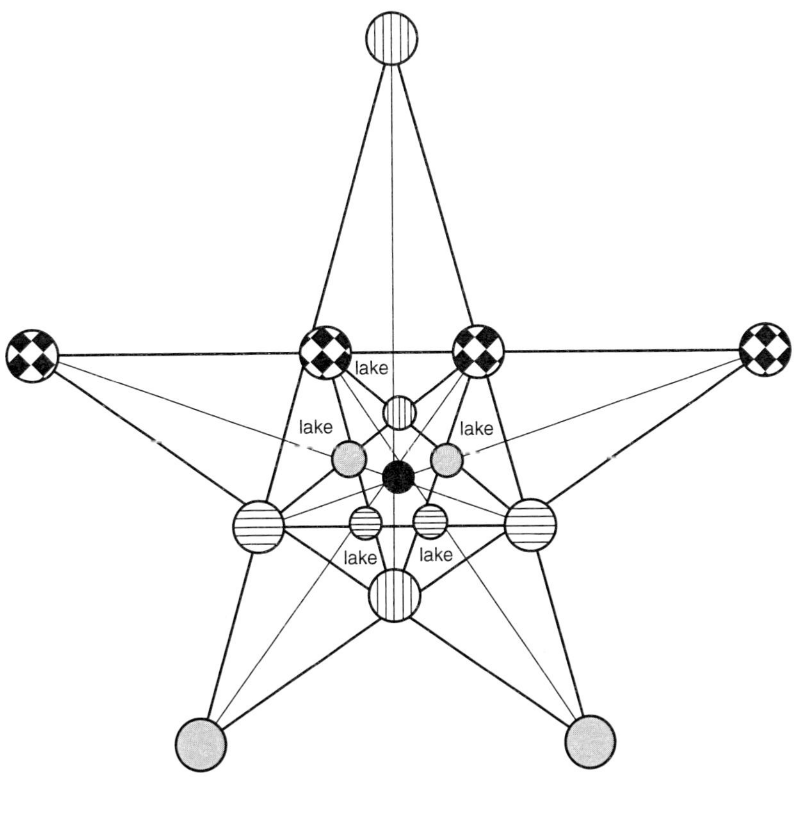
lake
lake
lake
lake
lake

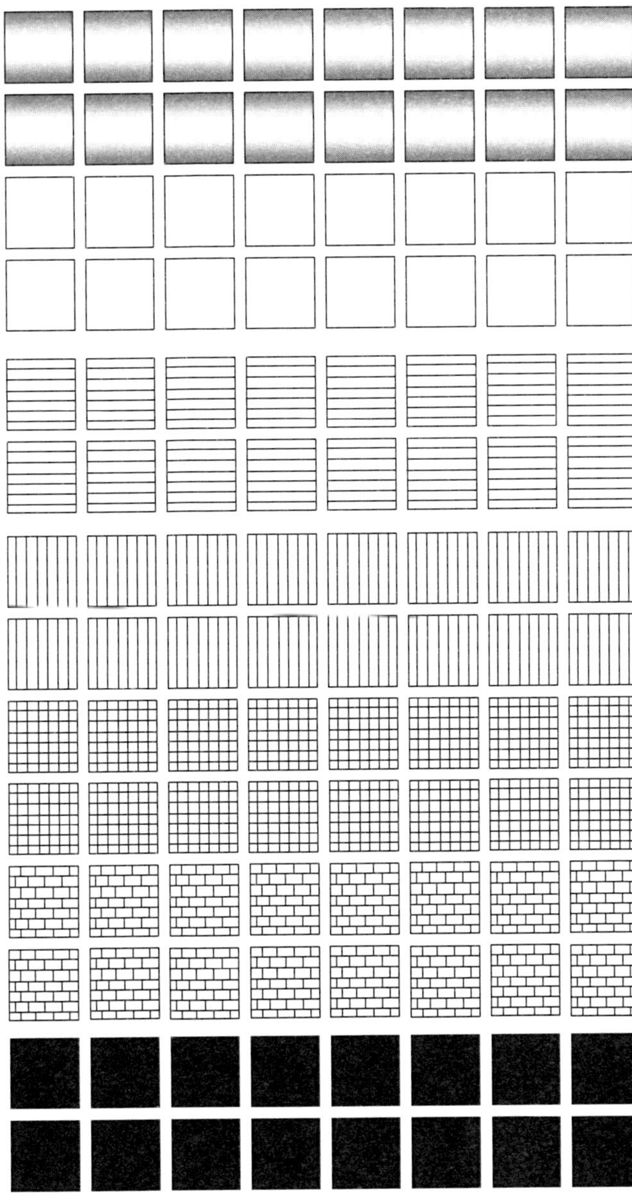

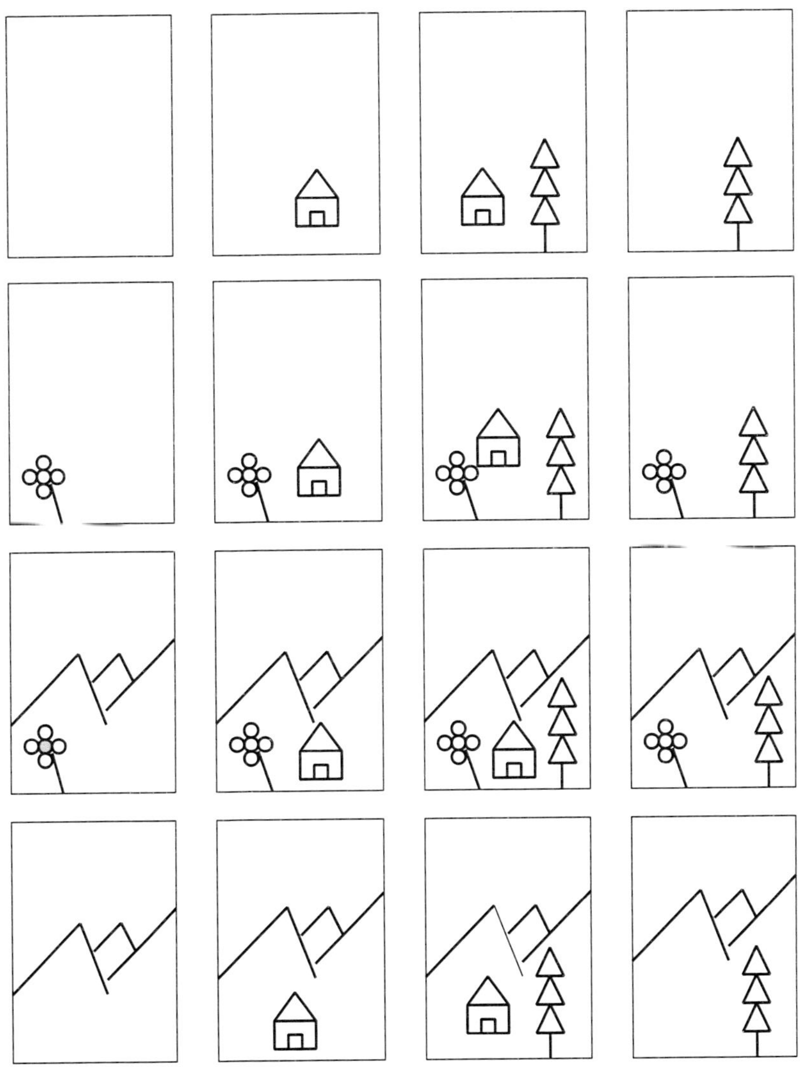

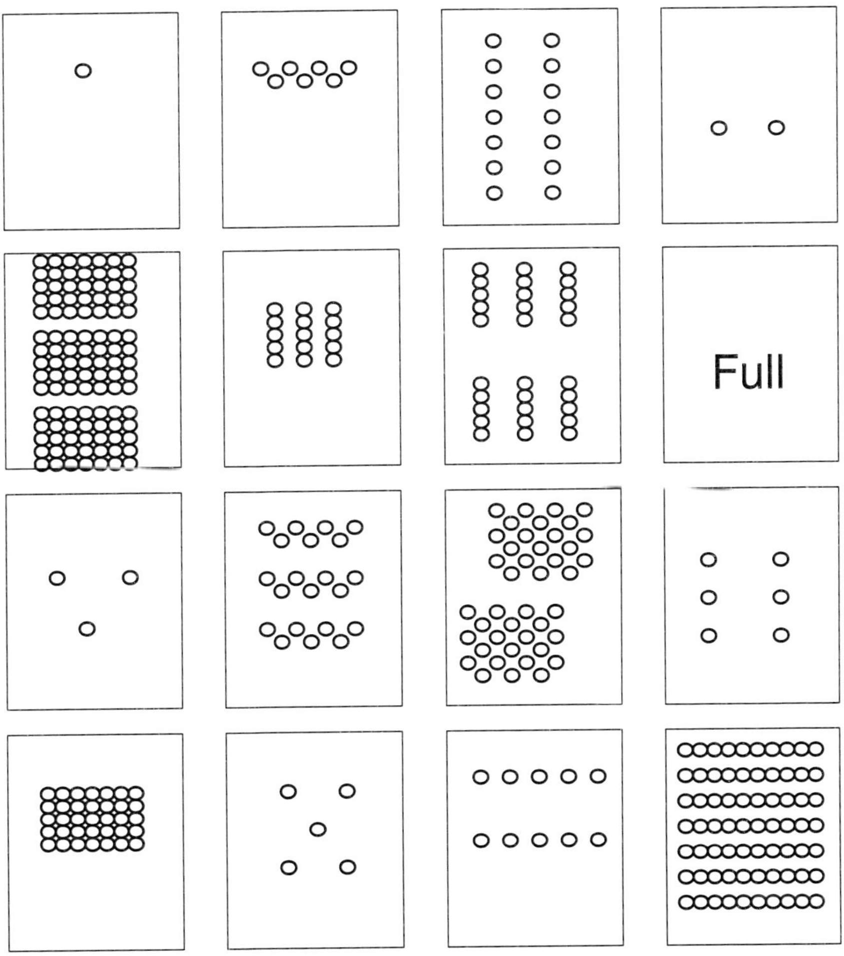
Full

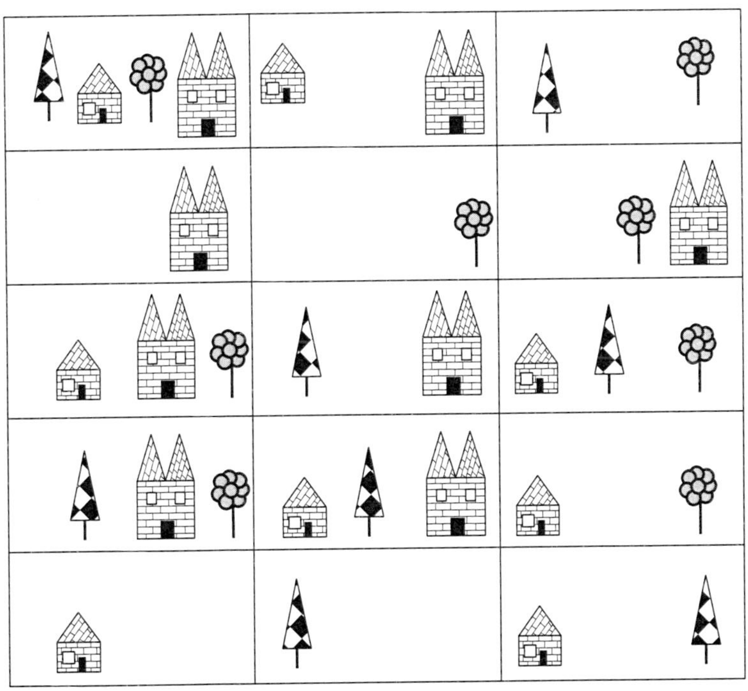

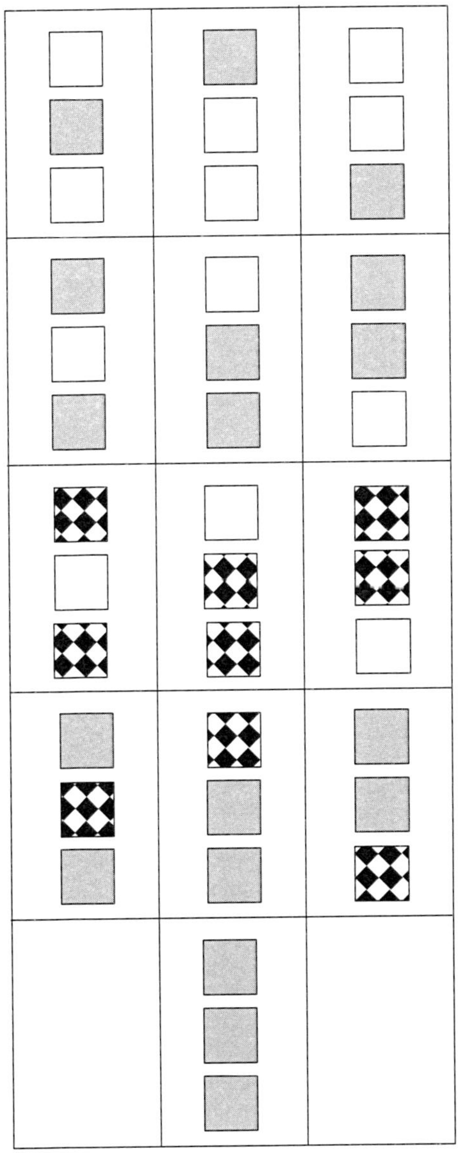

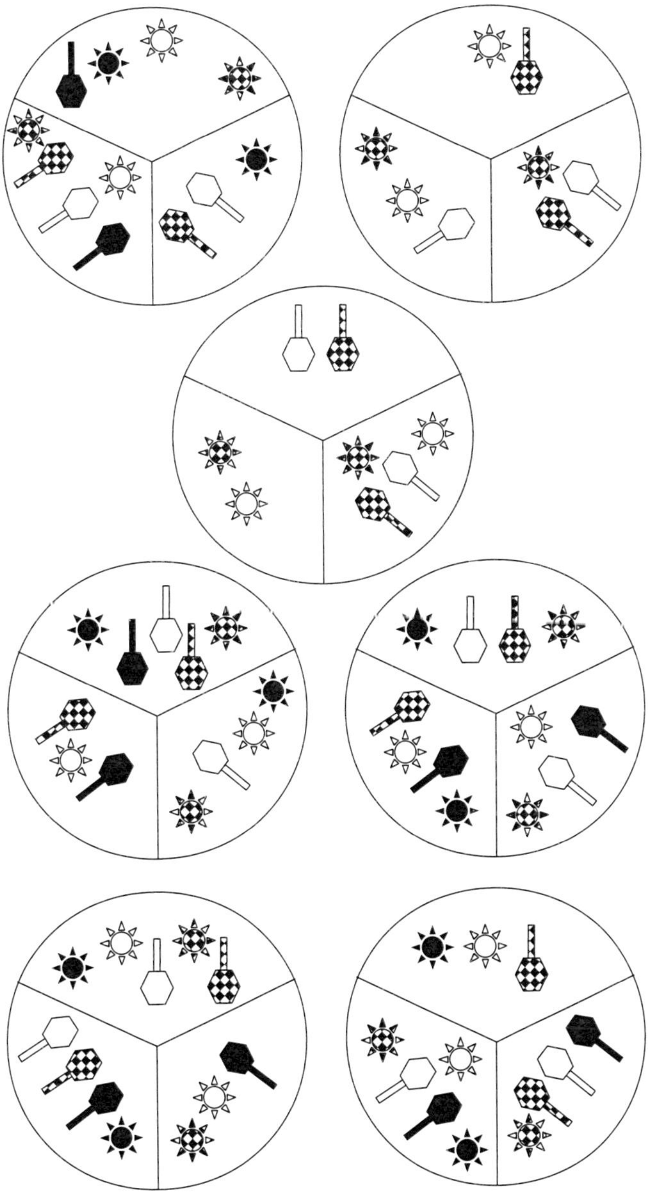

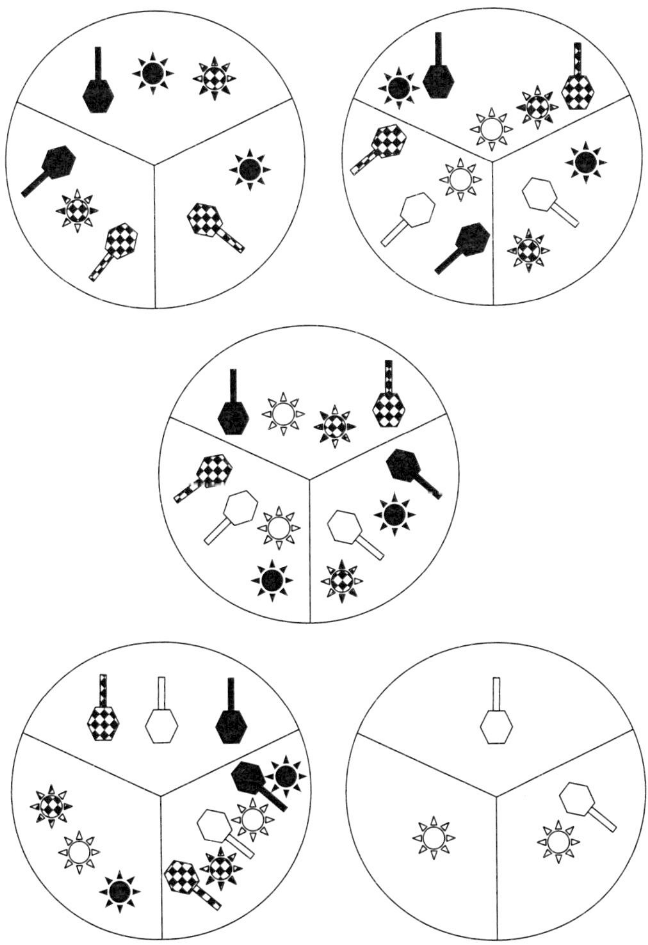

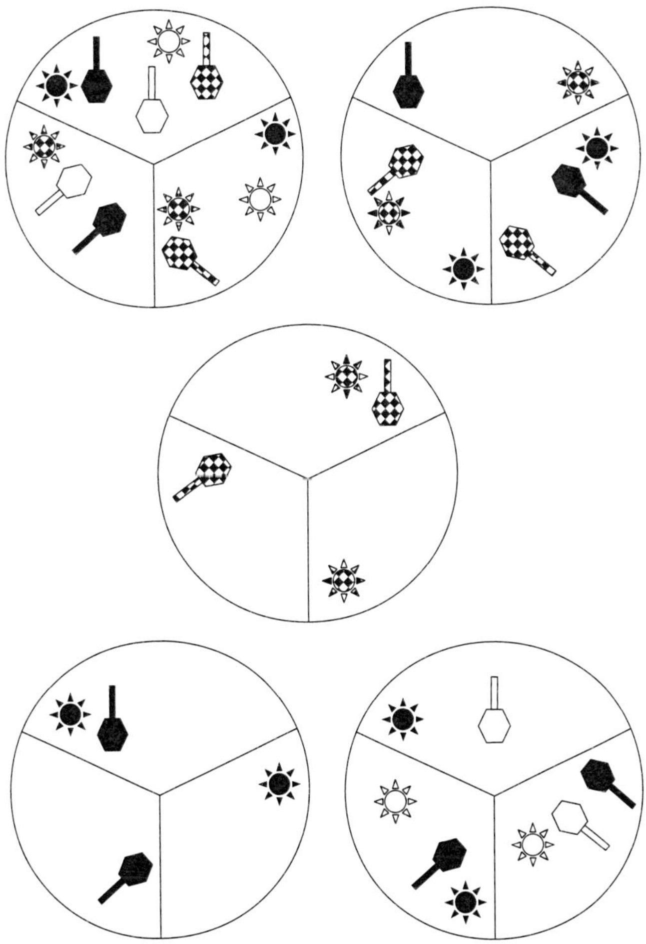

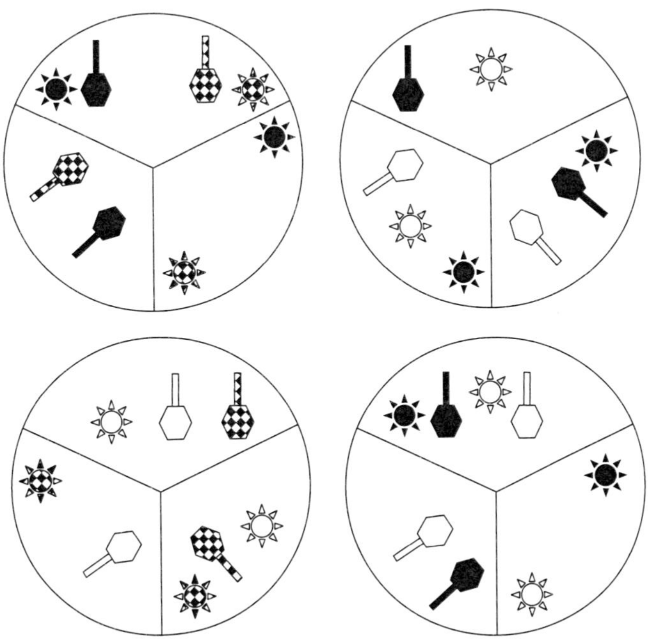

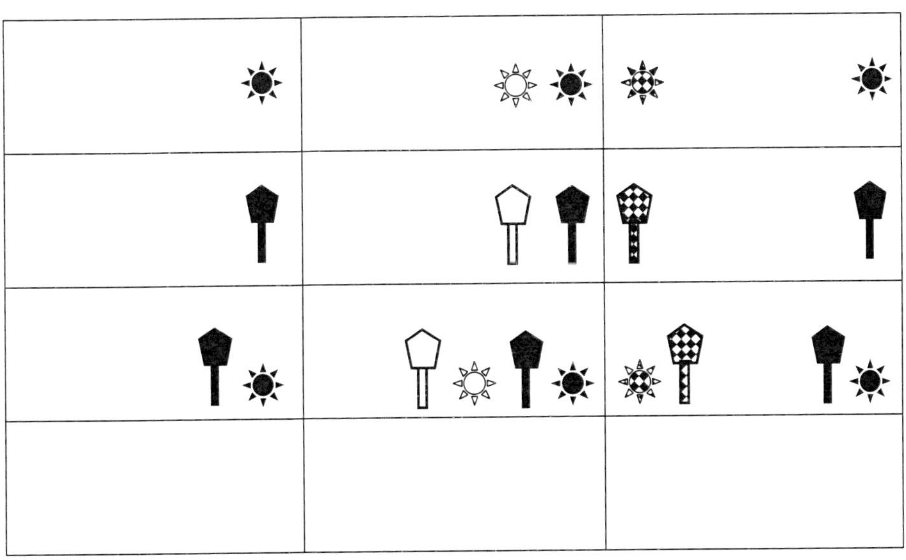

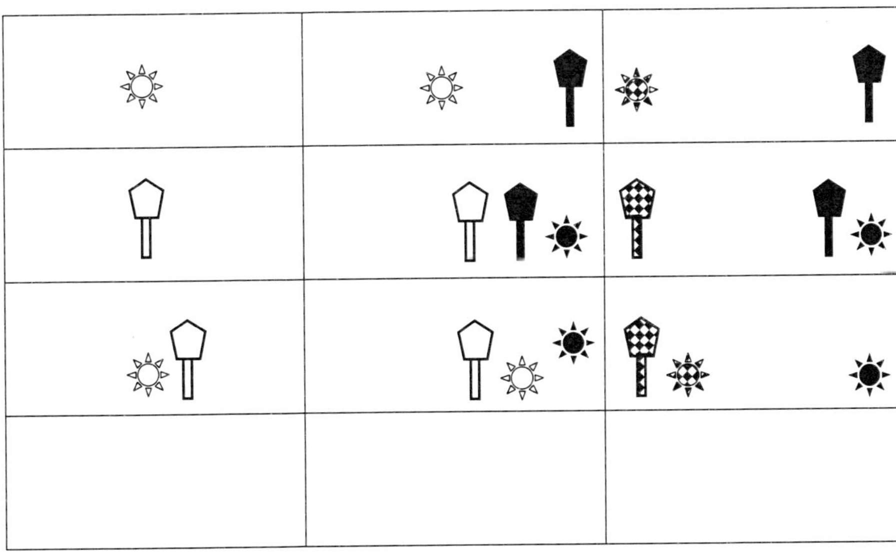

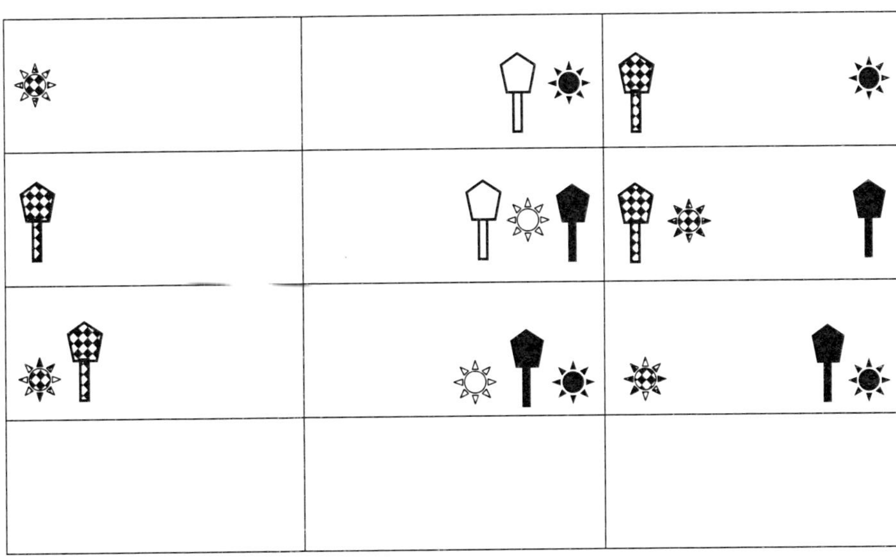

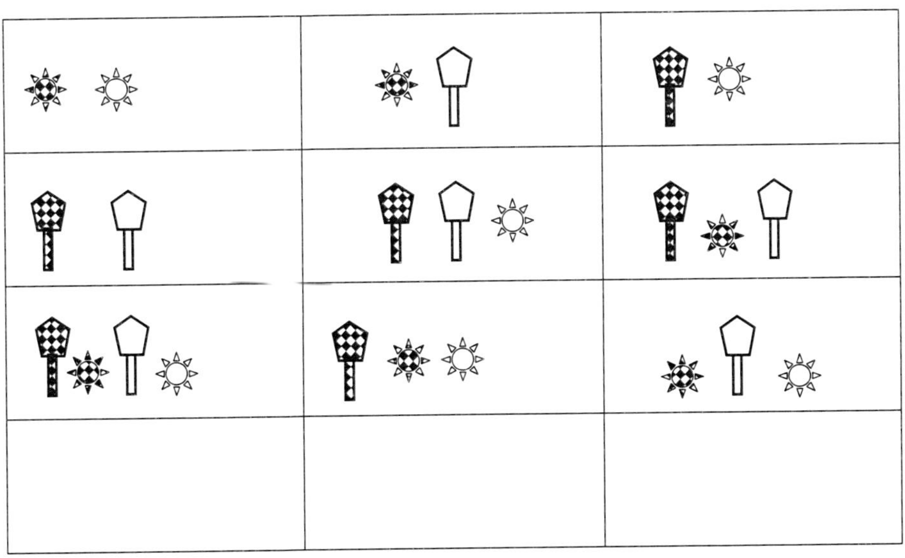

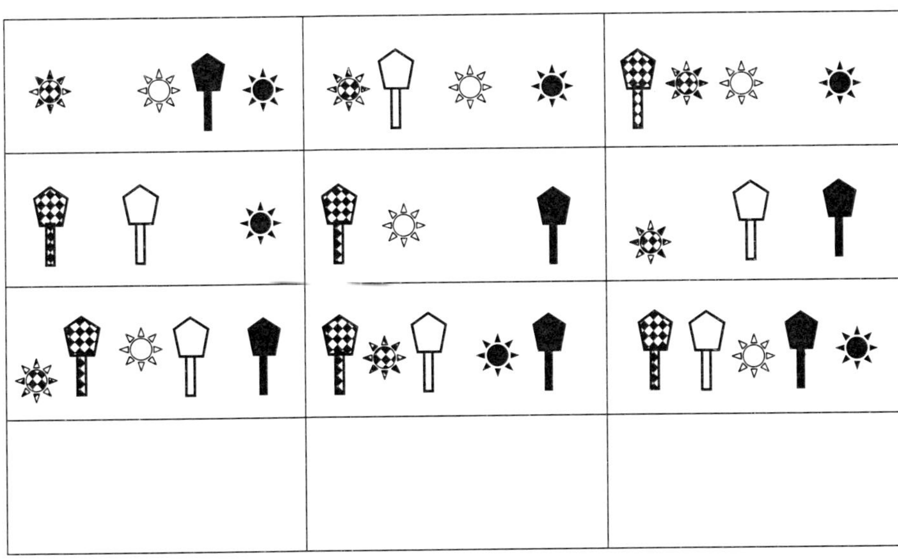

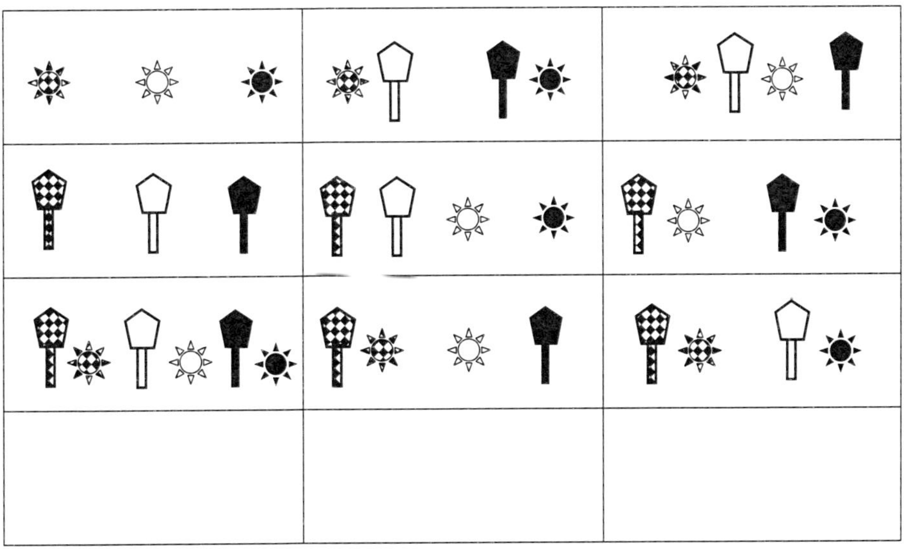

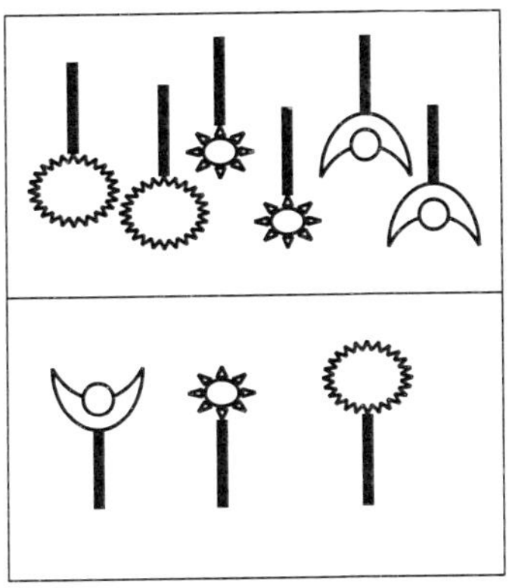
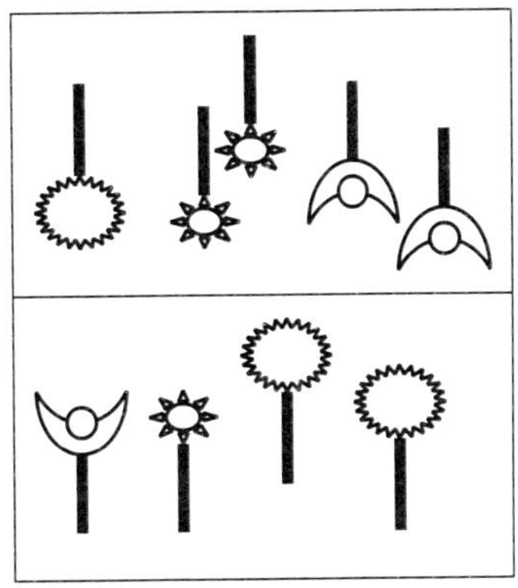

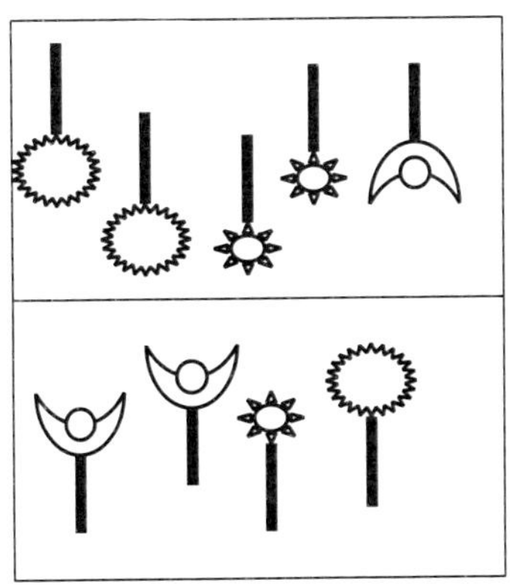

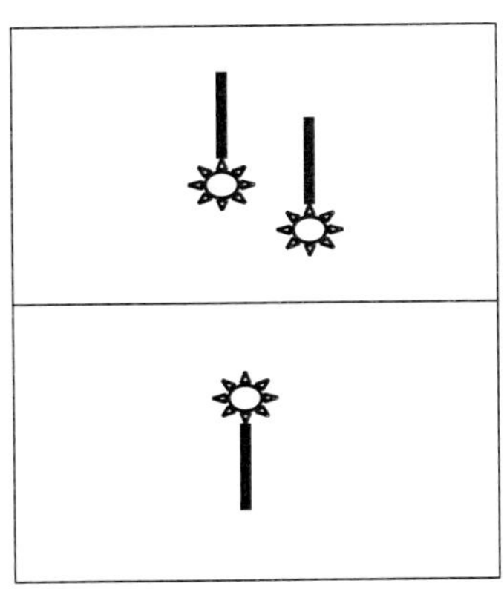

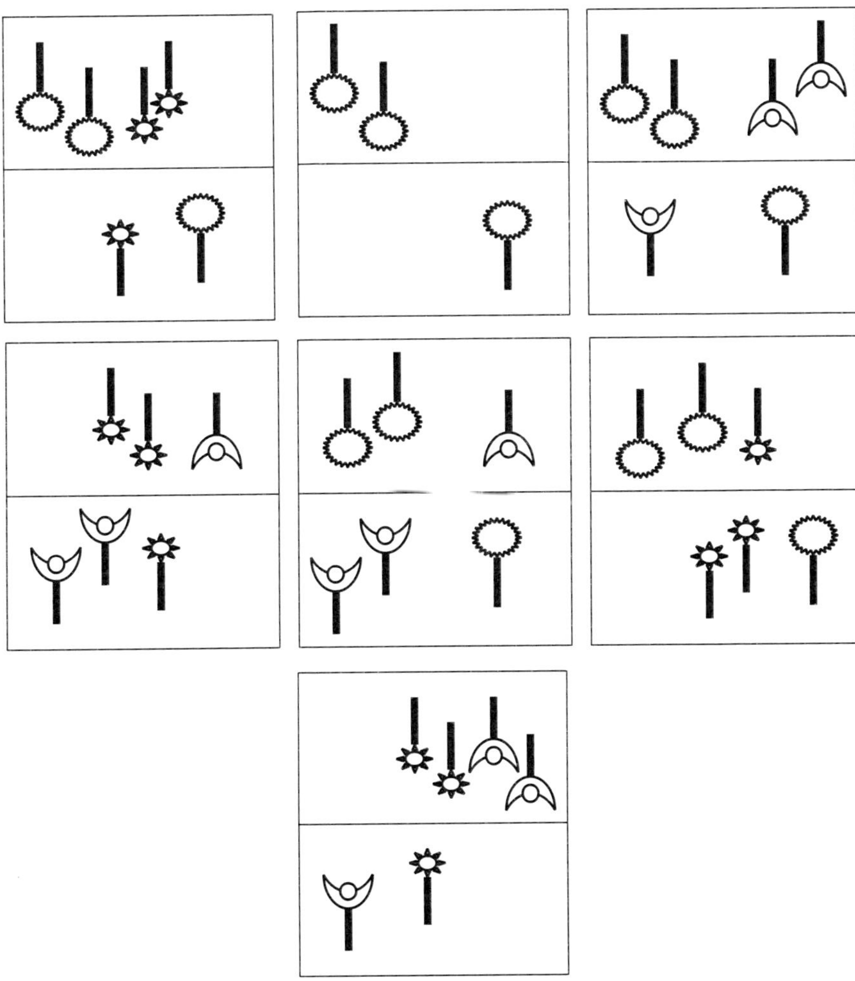

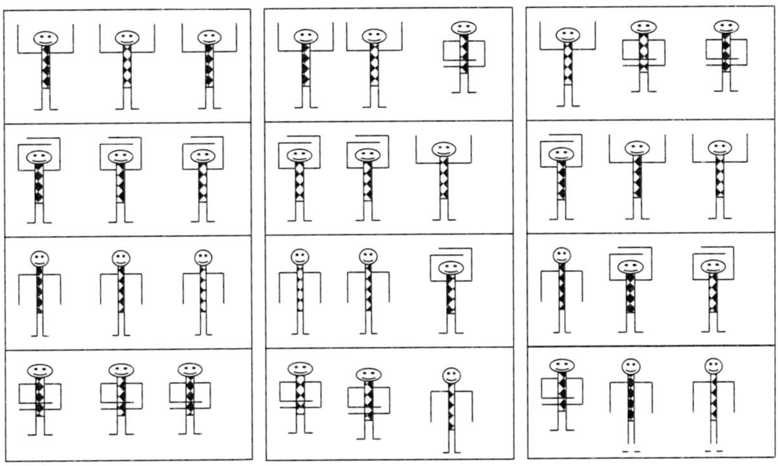

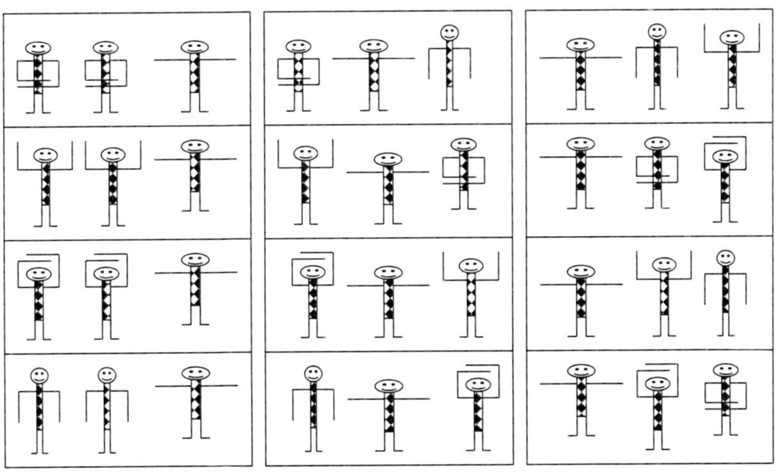

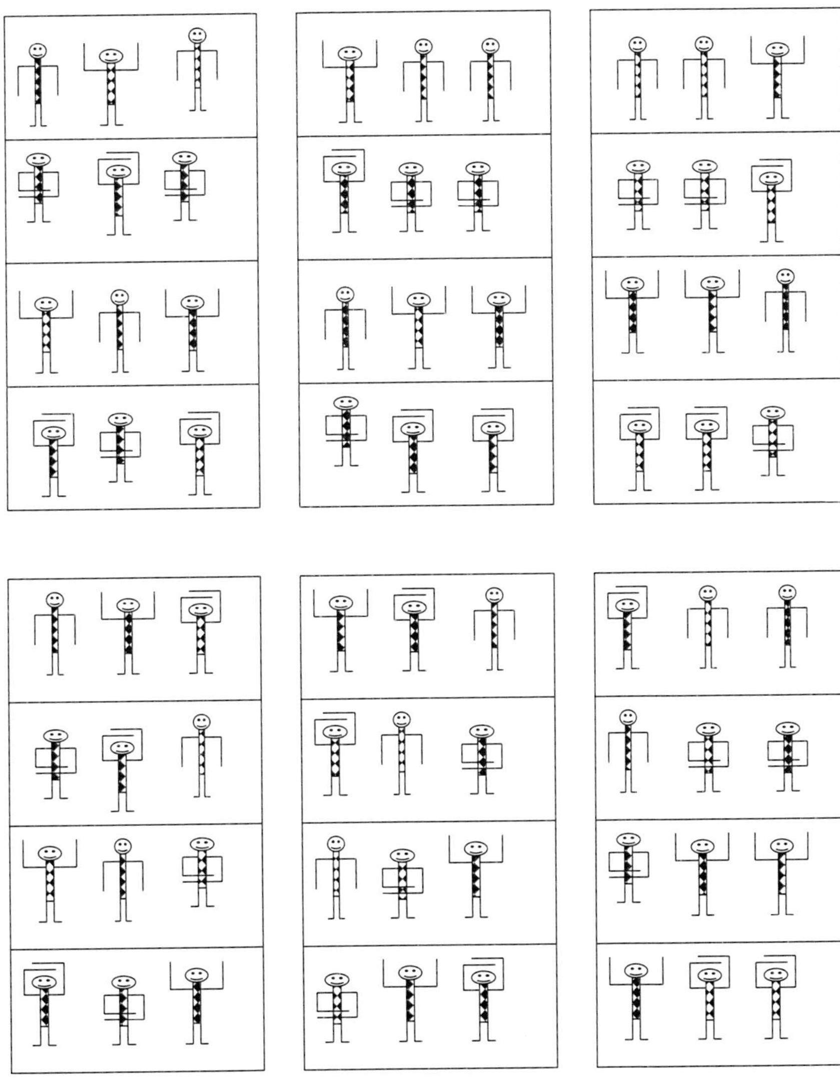

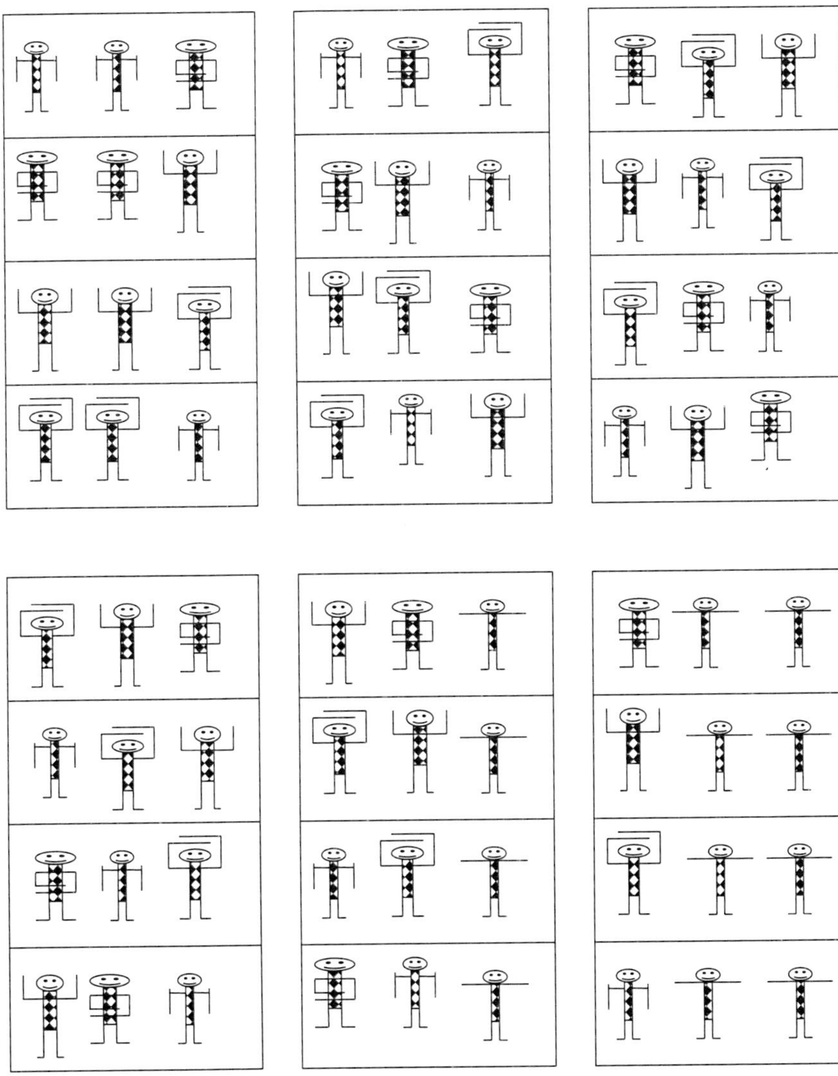

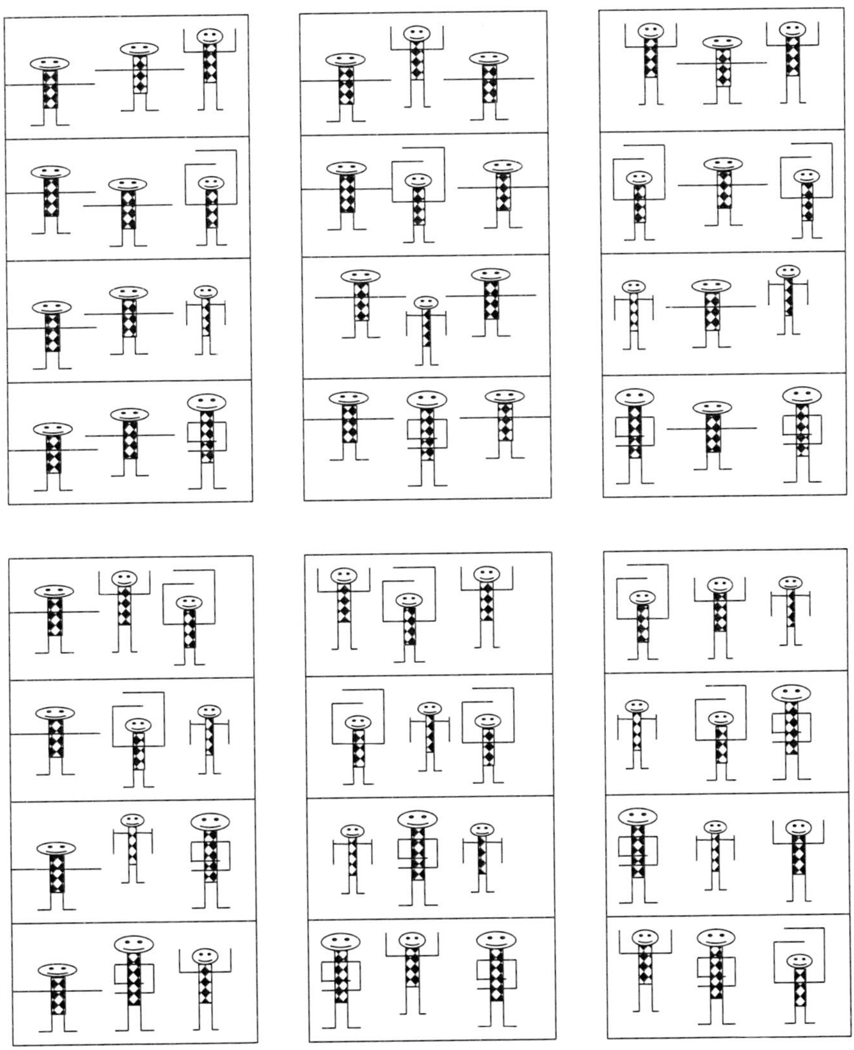

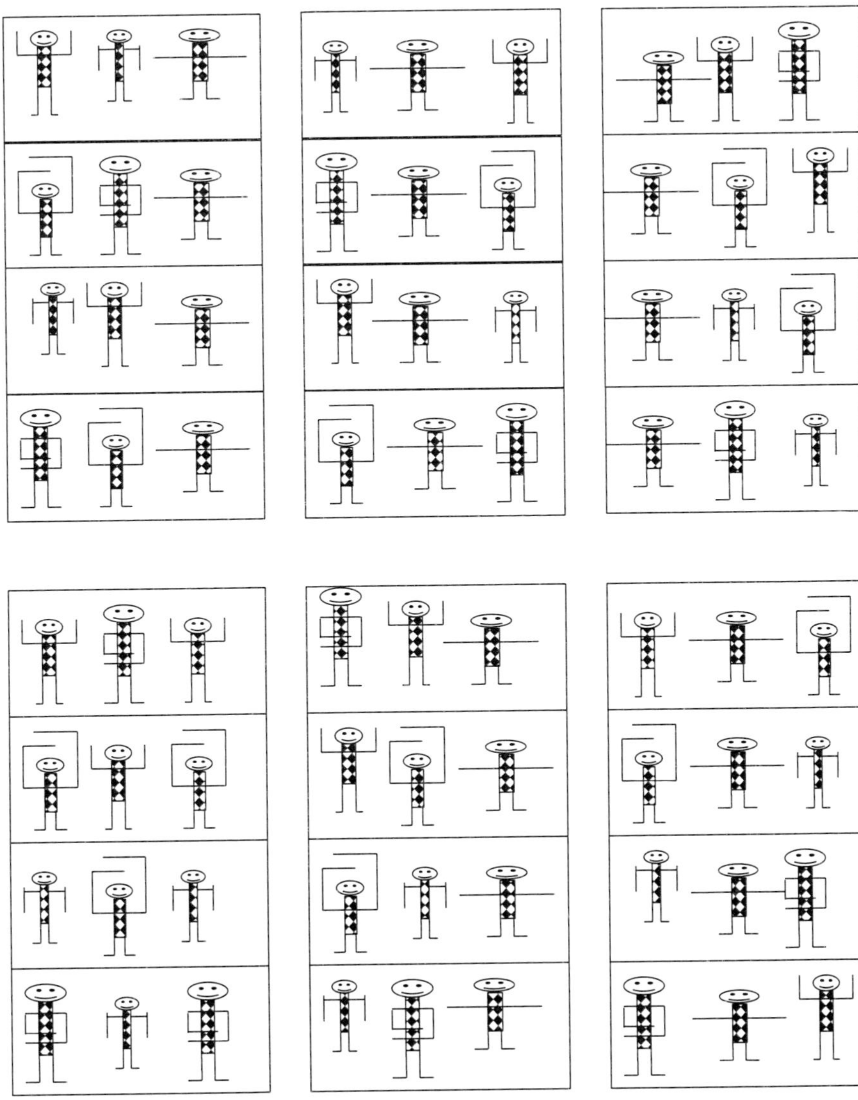

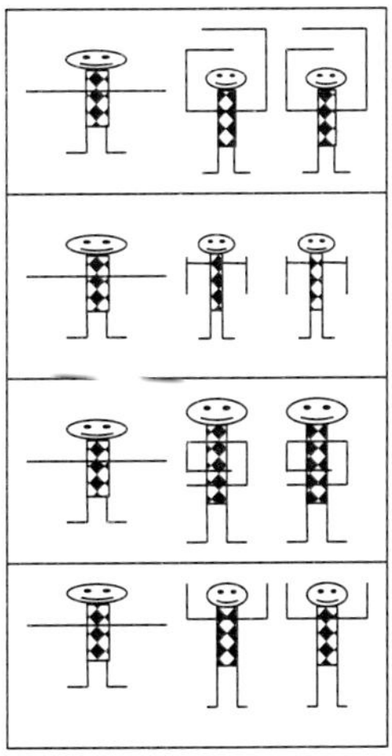